Beyond Shelter after Disaster: Practice, Process and Possibilities

Providing shelter after a disaster is recognised as one of the most complex areas of humanitarian relief and recovery. Some aid agencies have stopped providing shelter altogether after bad experiences, while those that do quickly become engaged in challenges that go far beyond the provision of structures alone. Yet with the number and severity of disasters set to increase, due to climate change and rapid urban growth, the need for approaches that work has never been greater.

This book explores the issues in three parts. The first, Practice, looks at lessons from past efforts. Part two, Process, proposes practical and effective people-centred approaches. Part three considers currently neglected issues such as disability, human rights and urban-oriented approaches. Through practical case studies and academic research, *Beyond Shelter after Disaster* critiques past methods and explores future options for improving practice in one of the most complex areas of post disaster relief and recovery.

This book was originally published as a special issue in *Environmental Hazards: Human and Policy Dimensions.*

David Sanderson has over 20 years' experience in development and emergency practice, working with NGOs and donors. He is currently a Professor and Director of the Centre for Development and Emergency Practice (CENDEP) at Oxford Brookes University. He is a member of several advisory boards, teaches regularly in Madrid and Lugano, Switzerland, and is a Visiting Professor at Université Paris-Est Créteil, France.

Jeni Burnell trained as an architect before pursuing a career in community art, architecture and development. She has a Master's degree in development and emergency practice and is currently a Research Associate at CENDEP, Oxford Brookes University. Her specialisations include shelter after disaster and community-led development using the 'Small Change' approach.

Beyond Shelter after Disaster: Practice, Process and Possibilities

With a foreword by
Edmund Penning-Roswell

Edited by
David Sanderson and Jeni Burnell

Routledge
Taylor & Francis Group
LONDON AND NEW YORK

First published in paperback 2024

First published 2013
by Routledge
4 Park Square, Milton Park, Abingdon, Oxon OX14 4RN

and by Routledge
605 Third Avenue, New York, NY 10158

Routledge is an imprint of the Taylor & Francis Group, an informa business

This book is a reproduction of *Environmental Hazards: Human and Policy Dimensions*, vol. 10, issue 3-4. The Publisher requests to those authors who may be citing this book to state, also, the bibliographical details of the special issue on which the book was based.

Publisher's Note
The publisher would like to make readers aware that the chapters in this book may be referred to as articles as they are identical to the articles published in the special issue. The publisher accepts responsibility for any inconsistencies that may have arisen in the course of preparing this volume for print.

British Library Cataloguing in Publication Data
A catalogue record for this book is available from the British Library

ISBN: 978-0-415-81423-2 (hbk)
ISBN: 978-1-03-292943-9 (pbk)
ISBN: 978-1-315-87184-4 (ebk)

DOI: 10.4324/9781315871844

Typeset in Helvetica
by Taylor & Francis Books

Contents

Citation Information

The chapters in this book were originally published in the *Environmental Hazards: Human and Policy Dimensions*, volume 10, issue 3-4 (July 2011). When citing this material, please use the original page numbering for each article, as follows:

Chapter 1
Whose reality counts?: Shelter after disaster
Jeni Burnell and David Sanderson
Environmental Hazards: Human and Policy Dimensions, volume 10, issue 3-4 (July 2011) pp. 189-192

Chapter 2
What have we learned from 40 years' experience of Disaster Shelter?
Ian Davis
Environmental Hazards: Human and Policy Dimensions, volume 10, issue 3-4 (July 2011) pp. 193-212

Chapter 3
Harnessing time: Reflections on constraints to development
Michal Lyons, Theo Schilderman and David Sanderson
Environmental Hazards: Human and Policy Dimensions, volume 10, issue 3-4 (July 2011) pp. 213-217

Chapter 4
Post-disaster reconstruction: A current analysis of Gujarat's response after the 2001 earthquake
Plato Jack Powell
Environmental Hazards: Human and Policy Dimensions, volume 10, issue 3-4 (July 2011) pp. 279-292

Chapter 5
Resilient dwellings or resilient people? Towards people-centred reconstruction
Theo Schilderman and Michal Lyons
Environmental Hazards: Human and Policy Dimensions, volume 10, issue 3-4 (July 2011) pp. 218-231

Chapter 6

Community-led resettlement: From a flood-affected slum to a new society in Pune, India
Victoria Cronin and Peter Guthrie
Environmental Hazards: Human and Policy Dimensions, volume 10, issue 3-4 (July 2011) pp. 310-326

Chapter 7

Lost in translation? The challenges of an equitable post-disaster reconstruction process: Lessons from Chile
Camillo Boano and Marisol García
Environmental Hazards: Human and Policy Dimensions, volume 10, issue 3-4 (July 2011) pp. 293-309

Chapter 8

Is there a human right to shelter after disaster?
Richard Carver
Environmental Hazards: Human and Policy Dimensions, volume 10, issue 3-4 (July 2011) pp. 232-247

Chapter 9

Can humanitarian responses in urban areas reinforce underlying causes of vulnerability? Tweaking a livelihoods analysis of inequality and infrastructure in splintering cities
Catherine Anna Crawford
Environmental Hazards: Human and Policy Dimensions, volume 10, issue 3-4 (July 2011) pp. 327-345

Chapter 10

From research to practice (and vice versa) for post-disaster settlement and shelter
Ilan Kelman, Joseph Ashmore, Esteban Leon and Sandra D'Urzo
Environmental Hazards: Human and Policy Dimensions, volume 10, issue 3-4 (July 2011) pp. 262-278

Chapter 11

Disability and public shelter in emergencies
John Twigg, Maria Kett, Helen Bottomley, Lin Tze Tan and Hussam Nasreddin
Environmental Hazards: Human and Policy Dimensions, volume 10, issue 3-4 (July 2011) pp. 248-261

Foreword

This volume discusses a subject of great importance. Decent housing – although different in different societies – is vital for human dignity, social welfare and family cohesion. Housing after disaster is even more crucial, in protecting the vulnerable and promoting recovery, not least by providing safety and security.

Shelter is the first step towards the restoration of this housing and security following a disaster, and is of great significance because without it recovery is impeded and the adverse effects of disaster are therefore extended in time and probably in severity. In showing the current deficiencies in terms of policy and practice this volume highlights the dilemmas for both governments and aid organisations. It is clear that shelter is vitally important, but it is often unclear why this shelter is not provided more quickly and with greater efficiency. We have learned a great deal about how the process of providing shelter can be improved, but it is not always the case – and sometimes far from it – that good practice is followed.

Let us hope that this volume contributes to the debate about improving practice, and thereby indirectly helps those afflicted by disasters and exposed to the calamities that life without shelter can bring. I warmly applaud the efforts of the authors of the relevant papers and the editorial team involved.

Edmund Penning-Rowsell
October 2012

Introduction

Beyond Shelter after disaster: practice, process and possibilities

JENI BURNELL AND DAVID SANDERSON

Centre for Development and Emergency Practice (CENDEP), School of the Built Environment, Oxford Brookes University, Gipsy Lane, Oxford OX3 0BP, UK

To say that the provision of shelter after disaster by humanitarian aid actors is costly, complicated and fraught with problems would probably be one of the greatest understatements in humanitarian aid. While often consuming large amounts of expenditure, too often the story of shelter after disaster has been an unhappy one. This year's Humanitarian Emergency Response Review (HERR), commissioned by the UK Government's Department for International Development (DFID, 2011, p. 25) to review how the UK Government should respond better to disasters, reached the same conclusion: 'Providing adequate shelter is one of the most intractable problems in international humanitarian response'. Reasons for this include sorting land ownership, materials procurement, organizing engagement with those affected by disaster and, increasingly for urban areas, density, rubble clearance, governmental involvement (or the lack of it), space and coordination. This last point especially was singled out by the HERR as not working well in relation to Emergency Shelter Cluster[1] coordination.

To these ends, this Special Edition of *Environmental Hazards: Human and Policy Dimensions* aims to examine the complexity of enacting good post-disaster shelter programmes. The papers in this edition result from a 2-day conference, *Improving Learning and Practice in the NGO Shelter Sector*, hosted by Oxford Brookes University's Centre for Development and Emergency Practice (CENDEP) in September 2010. Over 70 practitioners, academics – and sceptics – attended the event and covered a wide range of issues; some of the issues covered are reflected in the papers we have selected for this edition and range from urbanization, human rights and professionalization to construction, disability and theoretical understandings.

What though is 'shelter after disaster'? For this edition, the term refers to temporary structures beyond tents, and efforts by aid agencies to construct permanent post-disaster housing. The HERR refers to 'transitional shelter' – a broad term for housing that is something better than a tent but is not (intended to be a) permanent structure, usually designed with an intended lifespan of between 3 and 5 years. Transitional shelter has become the 'response of choice' for nearly all large aid agencies in recent years. Transitional shelter, however, is now raising questions, with critics arguing that it too often becomes permanent, that it uses up valuable aid resources and that it spends political will and donors' cash on short-term solutions that do little for addressing long-term problems.

In Haiti, following the devastating January 2010 earthquake, there have been real difficulties in implementing transitional shelter. Underpinning this has been Haiti's intractably difficult conditions of poverty and governance that existed before the earthquake. At least one large agency cancelled its programme outright, while others have found progress to be extremely slow. An independent study commissioned by the UK Disasters Emergency Committee (DEC) identified lessons learnt in the first year of operations of 13 of its member agencies, recommending that, 'In recovery, prioritise the facilitation of long term homes over the building of short term transitional shelters' (Clermont et al., 2011, p. 15), that is, that in effect agencies should rethink their engagement in transitional shelter. The authors of the report arrived at this conclusion after finding almost no support, within agencies or elsewhere, for transitional shelter as a useful solution – 'Many informants, both DEC member staff and others, have been very critical, labelling (transitional) shelters varyingly as "a total waste of money", "counter-developmental" and "suiting NGO timeframes and marketing needs" rather than people's needs' (Clermont et al., 2011, p. 15).

At the time of writing, some 800,000 people in Haiti are still living in tents, with solutions to the question of what to do still being seemingly far-off. An approach of the Government of Haiti to the resettlement of large numbers of people to new settlements, such as Corail, located several miles out of town, has been criticized as creating new, unserviced pockets of poverty. Strong among the critics is UN HABITAT, the United Nations agency tasked with urbanism and who at the time of writing lead the Emergency Shelter Cluster. UN HABITAT advocates the 'safe return' of affected dwellers to their original location (if it is safe to do so). Yet this also has its problems: if before the earthquake you lived in a four-storey block as a tenant, where, and what, do you return to? Also, if you were a squatter, or a renter, then what rights do you have?

Ian Davis opens this Special Edition with a candid reflection of his 40 years working in post-disaster reconstruction. As the author of the seminal 1978 publication 'Shelter after Disaster', Davis remains one of the key figures defining this field of work. Presented as a series of good practice principles, Davis's paper combines his experiences with other shelter practitioners. A key theme highlighted within his paper, reflected in many years of experience, is that 'relief is the enemy of recovery, so minimise relief, to maximise recovery' (Davis, this issue, p. 194). As the shelter sector grapples with transitional and other technological shelter solutions, it is imperative that assisting people to recover must remain at the centre of good shelter practice. This is a lesson still being learnt 40 years on.

How then do shelter professionals improve people's lives and living environments without creating dependence and increased vulnerability? Richard Carver's paper explores the issue from a human rights perspective, and asks, 'Is there a human right to shelter after disaster?' By beginning with the idea that, in legal terms, the answer is yes, his discussion explores what this means and crucially whether it is helpful for better practice. Discussing the seven criteria for adequate housing provided by the Committee for Economic, Social and Cultural Rights, and citing instances from Haiti and the tsunami, Carver asserts that the human rights approach becomes particularly valuable 'in the transition from the provision of shelter as an immediate emergency response into longer term reconstruction or development' (Carver, this issue, p. 243).

A key question that has emerged since the Haiti earthquake (and which was central to the DEC Haiti study) is 'What can be learnt for the next urban disaster?' Since 2010, the issue of improving disaster response in urban areas has been the subject of several reports, journals and conferences. Already a growing amount of work is taking place in this area, such as the Inter-Agency Standing Committee's (IASC) recently completed 'Strategy for meeting humanitarian challenges in urban areas' which identifies strategic objectives for making humanitarian responses in urban areas more effective (ISAC, 2010). As part of this work, IASC has identified some 90 tools and approaches for urban areas, applicable to various themes and sectors, not just for shelter. This work is in recognition that almost all of the aid communities' tools and understandings are rurally derived – we still talk of working 'in the field', even if it is a densely packed neighbourhood!

If UN HABITAT is right, then towns and cities are growing by some 1 million people a week. In post-disaster urban areas, shelter problems are compounded: density, confusion over land ownership, high-rise living and the presence of squatter settlements, often on poor-quality land, are just a few of the hurdles. Within this context, disasters are not only the shocks that occur after a sudden onset event, such as an earthquake, but, on a smaller every day level, are also the daily stresses that impact the lives and increase the vulnerabilities of the urban poor. Definitions of shelter after disaster therefore extend to reconstruction and, at times resettlement, of people living in vulnerable situations, often on marginal land, in cities and towns. Victoria Cronin and Peter Guthrie's paper 'Community-led resettlement: From a flood-affected slum to a new society in Pune, India' explores this theme, providing a detailed insight into the challenges and complexities of dealing with shifting governance structures in the building of new houses for a community affected by disaster. Focusing on a community that was devastated by floods in 1997, Cronin and Guthrie describe the engagement of civil society organizations and, after initial positive engagement from senior governmental administrators and politicians, the frequent changes of personnel that led to a start and stop process that lasted for several years. While the eventual process took some 6 years before building began, Cronin and Guthrie conclude that the successful elements of the project 'were very much driven by a collaborative NGO [non-governmental organisation] and CBO [community based organisation] partnership which enabled the beneficiaries to be in control of their housing situation' (Cronin and Guthrie, this issue, p. 320).

Camillo Boano and Marisol García's paper focuses on the recovery in urban areas in Constitución, Chile following the February 2010 earthquake. Boano and García point out that while cities can provide safe areas, they are too often places of massively enhanced risk, through poorly built buildings, ineffective building codes implementation and sheer density. Boano and García's paper also explores the roles of architects in post-disaster recovery, and the shift needed by traditional architects away from being (to cite British architect Jeremy Till) the 'interpretive agent' to that of 'transformative agency'. As part of this new skill set Boano and García note the need for built environment professionals, including architects, to promote 'an understanding of the capacity of people to act', that is, to build genuine participation and empowerment of people in design processes that effect them. In this way, Boano and García are echoing developmental specialist Robert Chambers' long-standing question and challenge, 'Whose reality counts?' – the professional who comes and goes for a short period, or people themselves, whose lives are being played out in the very places that professionals seek to assist in?

To these ends, Boano and García are touching on the wider question of the skill sets required to work in the shelter sector. The need for professionalization within the shelter sector with competent professionals has gained traction in recent years, with increasing numbers of new training programmes, such as the launch in 2011 of Save the Children UK's Shelter Trainee Scheme and a new Postgraduate Certificate in Shelter after Disaster offered by CENDEP.

If sector individuals need to have improved skills and competencies, then how do aid agencies improve also? Kelman et al.'s paper 'From research to practice (and vice versa) for post-disaster settlement and shelter' point to an absence of a meta-analysis of studies in this area. They highlight, importantly, the lack of ability for too many agencies to learn, observing that 'aid agencies rarely analyse the impact of their post-disaster settlement and shelter projects' (Kelman et al., this issue, pp. 263–264). One element for improving practice explored in the paper is improving the links between research and practice. Through an analysis of several disaster shelter responses, the paper discusses pragmatic issues relating to media interest (and its decline as time goes on), location and funding timeframe. Concerning the latter, they discuss trade-offs between quality and quantity, citing Somalia as an example: 'In 2009, one project delivered improved but basic shelter with a limited lifetime to over 24,000 people. In 2008, a separate project had provided durable sites and services (not shelter materials), but for only 700 people over 2 years and with significantly higher project costs per person assisted' (Kelman et al., this issue, p. 272).

Continuing the urban theme, Kate Crawford's paper asks whether urban recovery interventions by agencies can change societal divisions of wealth and poverty, or merely reinforce them. She cites as an example the provision of shelter kits, where a requirement for such a kit is land ownership. To rethink conceptual understandings of working in urban areas, in particular relating to infrastructure, Crawford presents a critique of an interpretation of the sustainable livelihoods approach, arguing instead for more nuanced understanding that wrestles with the complexities that urban reconstruction presents.

Dealing with complex realities and those who are particularly vulnerable is explored in John Twigg et al.'s paper. Through a literature review, this paper explores the practice of assisting people with disabilities in accessing public emergency shelters. The findings highlight a general failure within many emergency facilities to meet the basic needs of disabled people during an emergency. Twigg et al. cite one specialist who observes, 'Most disaster response systems are designed for people who can walk, run, see, drive, read, hear, speak and quickly understand and respond to instructions and alerts' (Twigg et al., this issue, p. 252). Twigg et al., somewhat like Crawford, argue that greater understandings are needed in differentiating the term 'vulnerability' during humanitarian actions.

If temporary shelter is complex, then arguably more so is permanent housing, with similar challenges of land ownership, location, durability, safety and quality. Plato Jack Powell's paper examines the pitfalls of post-disaster reconstruction through a study of several housing projects in India's Gujarat State, 10 years after the 2001 earthquake. His field research compares the success (in terms of quality and appropriateness) of buildings which were constructed by either a donor or owner-driven approach. To elaborate, donor-driven reconstruction is a commonly used approach in which 'donors – including governments, multilateral or bilateral agencies or humanitarian agencies – decide how and what to building and construct this directly or through contractors' (Schilderman, 2010, p. 26). Conversely, owner-driven reconstruction requires that donors support people in the process of constructing their own housing in accordance with their personal needs and requirements. Powell's paper emphasizes the flaws in policy leading to unenforced safe building regulations, which resulted, in many cases, in people's vulnerabilities being increased instead of reduced.

Powell's paper, like many others written on the above issues, highlights the inadequacy of donor-driven reconstruction to effectively meet people's needs. Donor-driven housing can fail spectacularly to put people in the centre. This failure is provided by the example of 'World Vision Nagar', a post-tsunami housing project built in India Tamil Nadu State. World Vision Nagar, as the name implies, puts the NGO in the centre, not the people. This focus on the provider is emphasized further: the NGO's name is on every house (reinforcing to every dweller that their home is a gift from a generous benefactor), and most disturbing of all, next to the main entrance there is a billboard which depicts the wave that brought such devastation and a tearful woman. What a reminder of how the colony's loved ones were killed!

Fortunately, good examples of projects that focused on processes (not products) and people can be found. After the Gujarat earthquake the Indian NGO SEEDS India trained local masons who spend time in villages rebuilding houses with local communities. The process took longer, but the outcome – houses designed and built with people in mind using local practices – were appropriate, and cost roughly half the price for houses twice the size compared to other nearby NGO-built housing.

The approach adopted by SEEDS India might be termed by Theo Schilderman and Michal Lyons 'People-Centred Reconstruction' (PCR). Their paper challenges existing norms of providing shelter after disaster and offers an alternative that puts people in the centre of the reconstruction process with the aim of making people more resilient to future risks. To achieve this, Schilderman and Lyons outline key components for achieving PCR and advocate that this approach 'requires not just making their buildings safer to live and work in, but also make making people more capable to adapt to risk and change' (Schilderman and Lyons, this issue, p. 227). It is an approach that aims to tackle people's underlying vulnerabilities to disasters and, in so doing, to bridge the gap between relief and long-term development. For this to happen, both policy and practice need to be challenged. Schilderman, in a 1993 paper, 'Disaster and development: a case study from Peru', discussed this issue, concluding that NGOs have an important role to play at both the local and the international level. Schilderman sees NGOs as catalysts for developing innovative strategies for reducing people's vulnerabilities while having 'an even more important role at the international level, in challenging the assumptions underlying current relief and reconstruction work, and in stimulating a change towards a more development-orientated approach' (Schilderman, 1993, p. 423).

Shelter after disaster therefore is costly, complicated and fraught with problems, and currently too many shelter programmes fall short. Hopefully this Special Edition will provide some contribution to this complex area of work. If the reader takes away anything from this edition however, it should be that good practice in 'shelter after disaster' puts people first – after all, *whose reality counts?*

Note

1. For more information on the Emergency Shelter Cluster please refer to: www.humanitarianreform.org.

References

Clermont, C., Sanderson, D., Sharma, A. and Spraos, H., 2011. *Urban Disasters – Lessons from Haiti: Study of Member Agencies' Responses to the Earthquake in Port au Price, Haiti, January 2010*. Report for the Disasters Emergency Committee (DEC), DEC, London. Accessed 31 May 2011. www.dec.org.uk/download/856/DEC-Haiti-urban-study.pdf.

Department for International Development (DFID), 2011. *Humanitarian Emergency Response Review*, 28 March 2011. Accessed 31 May 2011. DFID, London. www.dfid.gov.uk/Documents/publications1/HERR.pdf.

Inter-agency Standing Committee (IASC), 2010. *Final Strategy for Meeting Humanitarian Challenges in Urban Areas*. IASC (MHCUA), Geneva. Accessed: 31 May 2011. www.citiesalliance.org/ca/sites/citiesalliance.org/files/CA_Images/IASC_Strategy_Meeting_Humanitarian_Challenges_in_Urban_Areas%5B1%5D.pdf.

Schilderman, T., 1993. Disasters and development: a case study from Peru. *Journal of International Development* 5(4). 415–423.

Schilderman, T., 2010. Putting people at the centre of reconstruction. *Building Back Better: Delivering People-Centred Housing Reconstruction at Scale*, M. Lyons, T. Schilderman and C. Boano (2010) (eds). Practical Action Publishing, Rugby. 26.

What have we learned from 40 years' experience of Disaster Shelter?

IAN DAVIS

Centre for Development and Emergency Practice (CENDEP), School of the Built Environment, Oxford Brookes University, Gipsy Lane, Oxford, OX3 0PB, UK

The following paper is an expanded version of a lecture given on the same topic in a conference on Disaster Shelter organized by the Centre for Development and Emergency Practice (CENDEP) at Oxford Brookes University in September 2010. The paper covers a clutch of related topics: shelter, reconstruction, safety and development. Its main focus is on developing countries, but much can apply to first-world disaster recovery situations.

1. Shelter: That intractable problem...

> *'What we have learned in one long year of mourning after Haiti's earthquake is that an exogenous plan of reconstruction- one that is profit driven, exclusionary, conceived of and implemented by non-Haitians – can not reconstruct Haiti. It is the solemn obligation of all Haitians to join in the reconstruction and to have a voice in the direction of the nation'* (Jean-Bertrand Aristide, Former President of Haiti (2011)).

> *'We have many problems and the situation we find ourselves in is dreadful ...things have gotten worse and worse, we do not have work and we do not have money. There is no supervision. We are shown hope, but nothing has come to us except the hurricane season. Must we wait for another disaster, when things are so difficult for us? What will be done for those of us living in tents? We are eating dust. We want to go home. How can you help?'*

> *'There are talks of a rebuilding process, but nothing has happened. Must we wait forever? We want to find work since it is very painful to be dependent on others for help'* (Amboise Fleuristil – writing a message to a post box set up in Haiti by the International Organisation for Migration (IOM) on the first anniversary of the earthquake of January 10 2010 (2011)).

> *'Providing adequate shelter is one of the most intractable problems in international humanitarian response. Tents are too costly and do not last long enough. Plastic sheeting can be good but most often is low quality and falls apart immediately. Rebuilding houses takes years, even when land issues are not major obstacles.'*

This last quote is the dismal conclusion from the recent 'Humanitarian Emergency Response Review' chaired by Lord (Paddy) Ashdown and echoes the sentiments from Haiti (Ashdown, 2011). But why is shelter so problematic? The commonly used acronym: 'SAD' (Shelter after Disaster) has proved to be apt. One explanation for the problems may result from a mismatch. This is the difference between the importance disaster survivors place on sheltering and the low value attached by most donors. Graham Saunders, Head of Shelter and Settlements in the International Federation of Red Cross Societies (IFRC), states that the sector has failed to develop because the institutions have not advanced their own understanding of the subject despite progress at field levels (Saunders, 2011). Thus, out of the 472 National and International Humanitarian Agency signatories of the Code of Conduct for the International Red Cross and Red Crescent Movement in Disaster Relief, a mere dozen employ full time expert advisors with experience in shelter or housing reconstruction.

When challenging agencies about this neglect, most indicate their preference to work in the safety of familiar areas – health, nutrition, water or sanitation. Shelter is a costly commitment in time, energy and cash leading in a seamless manner straight into the complexities of housing reconstruction. This risky, yet vitally important world, remains an area where few external agencies wish to enter. Shelter and housing reconstruction is perceived to be too expensive with protracted involvement and working closely with national governments risks entanglement in red tape.

In 1972, I embarked on a study of shelter following disaster. This preoccupation with this elusive, problematic and expanding subject continues almost 40 years later, and this paper provides the first opportunity to reflect on the cumulative lessons.

It all began as PhD research in the Development Planning Unit of University College, London (Davis, 1985). I was fortunate to have my research supervised by a world authority in low-cost housing in developing countries, Professor Otto Koenigsberger. He had attached some demanding conditions to his agreement to oversee my research, namely that I would agree to visit the next major disaster to observe the shelter process, with 'eyes and ears open, mouth firmly shut and a firm promise not to design anything.'

But asking a lecturer to maintain silence is a rather fruitless cause and asking an architect to abstain from designing is akin to asking a newborn baby to avoid milk. Less than 3 months later, on 22 December 1972, Managua, the capital city of Nicaragua was devastated by an earthquake and shortly afterwards, with much naivety and trepidation, I arranged my travel to the ruined city.

Before I left Koenigsberger offered me some further perplexing advice:

> *'remember that relief is the enemy of recovery, so minimise relief, to maximise recovery.'*

I travelled to Managua via Columbus, Ohio to meet two of the founders of social science research into disasters: Henry Quarantelli and Russell Dynes. They generously spent 2 days showing me the extensive library of the Disaster Research Centre. Henry, drove me to the airport and as we parted at the departure gate he offered me some invaluable advice for my fieldwork:

> *'....when you get to the disaster site you will find that there will have been a gross exaggeration of damage and dislocation, and a massive underestimate of local resources to tackle their emerging problems.'*

The wisdom of this initial advice concerning the need to observe and listen, to recognize that design is not the exclusive preserve of professionals, to be aware of the dangers of dependency and the need to recognize the capacity of survivors have remained central lessons for over four decades (refer to 4.3. Third essential approach: build capacity). The risks of paternalism, duplication of effort, inaccurate damage and needs assessments and misplaced assistance have all expanded as unresolved issues at the same explosive pace as the subject itself.

The following paper is an expanded version of a lecture given on the same topic at the conference, *Improving Learning and Practice in the NGO Shelter Sector*, hosted by the Centre for Development and Emergency Practice (CENDEP), Oxford Brookes University in September 2010.

The paper covers a clutch of related topics: shelter, reconstruction, safety and development.

While its main focus is on developing countries, much can also apply to first world disaster recovery situations. The paper is structured in four parts; 1. Summary of progress with shelter is set out in three connected tables that relate to shelter policy and practice, safe reconstruction and the role of house users in the building process; 2. Expert opinion on shelter; 3. Common themes; and 4. Essential approaches to shelter.

1. Summary of progress 1972–2011

Progress with Shelter 1972–2011

- INTERTECT (Director Fred Cuny) provided consultancy support to Oxfam/US Government/UNHCR/etc. in varied refugee/ disaster relief/ reconstruction contexts 1968–1995.
- Krimgold, F., 1974. *Doctoral Thesis on Disaster Prevention* (with specific reference to Shelter issues). Royal Institute of Technology, Stockholm.
- Davis (1978). *Shelter after Disaster*. Oxford Polytechnic Press, Oxford.
  ☐ (First book on Disaster Shelter).
- UNDRO (1982). *Shelter after Disaster – Guidelines for Assistance*, I. Davis (ed.) with F. Cuny, F. Krimgold, A. Fernandez and P. Thompson. Office of UN Disaster Relief Coordinator (UNDRO), New York.
  ☐ (First set of UN Shelter Guidelines) These Guidelines are being 'refreshed' (UNOCHA, 2011).
- Aysan and Oliver (1987). *Housing and Culture after Earthquakes, A Guide for Policy Making on Housing in Seismic Areas.* Oxford Polytechnic, Oxford.
  ☐ (First guidance on cultural aspects of shelter and housing reconstruction).
- Sphere (2004). *The Sphere Project, The Humanitarian Charter and Minimum Standards in Disaster Response.* Sphere and Oxfam Publishing, Geneva.
  ☐ (Guidelines on Humanitarian Assistance for International NGO Sector included shelter performance standards from 1998. The Sphere Guidelines were issued in a revised edition in 2011 with updated shelter standards).
- From 1972 until about the time of the Gujarat Earthquake of 2001 and later the Asian Tsunami in 2004/5 there was a low level of interest in shelter/reconstruction by the UN and most NGO sector (Davis, 1981; Aysan and Davis, 1992; Davis, 2009).
- The Shelter Project, later to become the Sheltercentre, was a key influence in raising agency awareness from 2004 until the present. Advice and training is provided primarily to the international shelter community (Corsellis and Vitale, 2005).
- Sheltercentre (2010). *Shelter after Disaster: Strategies for Transitional Settlement and Reconstruction.* UNOCHA, Geneva.
  ☐ (replacement of the 'Shelter after Disaster' 1982 Guidelines).
- From 2005 Formation by UN of three Clusters with links to Emergency Shelter. *Emergency Shelter in Disaster Situations* (coordinator: IFRC); *Emergency Shelter in Areas of Conflict* (coordinator: UNHCR); *Early Recovery Cluster* (coordinator UNDP).[1]
- Abbas, K. Jah, A.K. with Barenstein, J.D., Phelps, P.M., Pettit, D. and Sena, S. (2009). *Safer Homes, Stronger Communities. A Handbook for Reconstruction*. The World Bank, GFDRR Washington, DC.
- Ashmore (2008 and 2009). *Shelter Projects 2008 and Shelter Projects 2010.* UN-HABITAT and IFRC, Geneva.
- UNOCHA (2011). *Shelter after Disaster- Guidelines for Assistance- (Refreshed Version)*, I. Davis (ed.), F. Krimgold and P. Thompson. UNOCHA, Geneva.
  ☐ (Refreshed Version of UN Shelter Guidelines with new case studies and updating of each chapter).
- Masters Courses (or Modules of Masters Courses) in '*Shelter and Settlement after Disasters*' in Copenhagen/Lund Universities with support from IFRC (started 2011) and also a Post Graduate Certificate in '*Shelter after Disaster*' in the Centre for Development and Emergency Practice (CENDEP) Oxford Brookes University (started 2011).
  ☐ These are the first specific masters courses on shelter to be initiated.

Progress with hazard-resistant traditional dwelling construction 1972–2011

- From 1966 support for safe indigenous, non-engineered building came from various organizations: the UK Government's Building Research Establishment, UNESCO and the University of Roorkee in India as well as from pioneering individuals: D. Daldy/N.N. Ambraseys/R. Razani/A.S. Arya/Fred Cuny/Robin Spence, etc.
- The NGO *Intermediate Technology* founded by Fritz Schumacher (now renamed: *Practical Action*) provided support to this movement from the early 1970s until the present in work currently led by Theo Schilderman (Lyons and Schilderman, 2010). This organization has developed building safety programmes in Peru and other locations.
- From the late 1960s John Norton and his colleagues formed the NGO: *Development Workshop*. This pioneering organization developed hazard resistant low-cost housing in South Asia and in West Africa up to the present.
- The campaign for building safety in traditional buildings was stimulated by the International Decade for Natural Hazard Reduction (IDNDR) 1990–1999.
- *Building for Safety* Project. Four books were developed by teams in Cambridge University and Oxford Polytechnic (Clayton and Davis, 1994; Aysan et al., 1995; Coburn et al., 1995; Dudley and Haaland, 1993).
- There are key chapters in the following classic study on the safety of indigenous building from seismic impact: Coburn, A. and Spence, R. 2002. *Earthquake Protection*, 2nd edn. Wiley, Chichester.
- The creation of the *International Strategy for Disaster Reduction* (ISDR) in 2000 provided additional stimulus for building safety.
- From 2000 the Nepal based NGO *National Society for Earthquake Technology* (NSET) led by Amod Dixit has played a major role in building safety in Asia.
- The impact of the 2004 South Asian Tsunami was decisive as a 'tipping point' in agency awareness of safety needs.
- *The Hyogo Framework for Action* (HFA) 2005–2015 provided the first internationally agreed framework for Disaster Risk Reduction (DRR).
- UN-HABITAT, IFRC and the World Bank's *Global Facility for Disaster Reduction and Recovery* (GFDR) now provide vital leadership roles in safe construction in traditional low-cost buildings and settlements.

User involvement in safe housing reconstruction 1972–2011

- John F.C. Turner and his colleagues promoted a strong and highly influential movement for people to create their own buildings from the late 1960s.
- In 1976, Oxfam/World Neighbors organized an innovative user-build safe-housing reconstruction project following the Guatemala earthquake. Fred Cuny of INTERTECT gave project advice.
- The approach was further pursued after the Andhra Pradesh cyclone in 1977, when INTERTECT formed *Appropriate Reconstruction Training Information Centre* (ARTIC).
- Housing Reconstruction in the 1980s and 1990s was primarily contractor based. However, there were notable exceptions such as reconstruction after the 1993 Indian Latur earthquake when EFICOR/Tearfund/DFID adopted a safe user-build approach to reconstruction in the village of Malkondji.
- UN-HABITAT and the World Bank have promoted large scale user-build approaches to reconstruction in successive reconstruction projects: (after 2001 Gujarat earthquake, the 2005 South Asia Tsunami and the 2005 Pakistan and 2010 Haiti earthquakes).
- IFRC (2010). *Owner-Driven Housing Reconstruction Guidelines*. IFRC, Geneva.
- In 2011, User-Build approaches linked to training in safe construction for builders are becoming the default reconstruction policy of international agencies despite commercial pressures for contractor-based policies.

2. Expert opinion on shelter

Following the 2010 lecture, to check whether my perceptions of the subject were shared by others, and to enhance the scope of this paper, I approached a number of friends who have all had a positive influence on my own understanding of these complex subjects from 1972 until the present. They came from 11 countries and have worked in a range of different fields: geography, law, physics, engineering, but the majority (13 of the 18) come from architectural/physical planning backgrounds. They have had extensive experience of researching, consulting, lecturing, writing, advocacy, activism and direct involvement within the fields of disaster planning and risk management. And collectively they have worked in the shelter/safe housing reconstruction field at local, national and international levels.

I asked each of them to answer the following question:

'What has been the most important lesson you have learned from your own experience of Disaster Shelter or Reconstruction of Dwellings?'

My initial intention was to edit each comment into a standard length, but on reviewing the rich insights they have enthusiastically shared, I decided to do full justice to them by reproducing them in full. At the conclusion of their answers, some common themes are extracted.

2.1. Support survivors as they provide their own shelter

'Funding varies wildly between disasters, and this is often the most critical determinant on what external assistance can be given to support people to help themselves. As an example in the tsunami in Aceh, there was sufficient funding ($10,000 US) to build houses for all affectees, whilst in the current Pakistan floods it is doubtful whether all affected families will receive a tent or a tarpaulin. In other locations with chronic emergencies (e.g. Somalia, Eastern Chad), there are insufficient funds to maintain even the most basic of shelter support.'

'As a result, one of the main messages in the original version of Shelter after Disaster (UNDRO 1982) (with slightly more elegant wording) remains true – that the primary actors from disaster to recovery are usually the affectees themselves.'

Joe Ashmore

Independent Consultant, Extensive work with UN-HABITAT, Editor of 'Shelter Projects 2007, 8 and 9' published by UN-HABITAT and IFRC (Ashmore, 2009a and 2009b)

2.2. Respond to local needs as they define them

'The most important lesson I have learned is that the process of creating disaster shelter and reconstruction of dwellings is far more complex than what external assistance providers are generally prepared to address. Although it seems a cliché to reference them, in fact the underlying cultural, political, economic and geographic realities govern what is possible – not what the capabilities, resources and preconceptions are of outside agencies. The further from the disaster, time wise, the more difficult it is to respond appropriately. Rapid provision of plastic sheeting is relatively easy compared to understanding the criteria for taking the next steps: appropriate site with acceptable land tenure, environmentally sustainable, linked to employment opportunities, acceptable shelter and site design, and reinforced social networks and processes.'

'The lesson is, therefore, that external aid providers need to maintain a humble posture as servant and to furnish their assistance as a vehicle to support the local needs as defined by them, not as a "deliverable".'

Paul Thompson

Architect/planner consultant working with UN Agencies 1975 – present, founder of 'INTERWORKS' Madison, Wisconsin, USA

2.3. Actions by the survivors in providing their own shelter needs is becoming imperative

'As the frequency of disasters and associated shelter risks and vulnerabilities increase, the majority of

households whose homes are damaged or destroyed by such disasters have to meet their own needs with limited or no assistance from their Government of the humanitarian community. The scale of post-disaster shelter need that is increasingly emerging is beyond the response capacity of institutional humanitarians, be they governmental or non-governmental.'

'To remain relevant, the humanitarian community needs to recognise that its "value-added" is in promoting and sharing information – acting as an enabler through fostering linkages between established key local or regional actors or institutions in both the public and private sectors. The notion that the role of technical sectors such as shelter and settlement is just to provide select technical solutions and expertise is a misnomer. Enabling the political, economic and social environment through coordination, networking, localisation and the use of cash, information technology and a wide variety of approaches is key to impacting on the ever-increasing scale of post-disaster housing need.'

Graham Saunders
Head: Shelter & Settlements Department, International Federation of Red Cross and Red Crescent Societies (IFRC)

2.4. But whose reality counts?

'The biggest lesson is the blindingly obvious one of involving people in decisions that affect them – a thought that so often seems to elude those tasked with housing reconstruction. Those that have survived disasters may have lost everything, and then too often suffer a further indignity of being re-housed in houses that may be just wrong, on location, culture, materials or quality – and too often with a patronising sign on the front saying it is a "gift".'

'"Chambers" challenging question, "whose reality counts?" ought to be writ large on every agency proposal or project planning worksheet.'

David Sanderson
Architect, Director of the Centre for Development and Emergency Practice (CENDEP) Oxford Brookes University, Oxford, UK

2.5. The significance of shelter or dwelling as 'home'

'Twenty-five years into this kind of work from urban development to disaster risk management and shelter in disasters, and I have a one half-glass empty and one half-glass full of learning. The former is we not only seem to reinvent the wheel every time a major event happens, we seem to turn it the wrong way too. The latter, shelter after disasters, only works when it is perceived and driven by a profound understanding that it is not only protection from the elements or somewhere for people to store their belongings. It works when programmes are driven by the realisation that shelter is the foundation block for people's recovery.'

'Nothing happens when people do not have a "home". A "house" doesn't offer much and could be easily abandoned if inappropriate. These are two radically different notions. I am still looking forward to the time when the term and concept of "home" finds its way into not just the language of post-disaster recovery but into programme design.'

Mo Hamza
Architect/ Planner. Independent Consultant/ Research/ Lecturer, Oxford, UK

2.6. Building risk reduction into recovery planning

'Recovery programming can (– and often does –) reinforce both disaster risk and poverty, while at the same time providing a potentially powerful platform to counteract them. Three of the various reasons for this situation are:

(a) Housing/settlement recovery is still mainly dealt with in terms of location and construction quality as compared to a broader understanding of housing and settlements with regard to their social, environmental, economic, political, and institutional aspects, significance, influences and connections to other issues.

(b) Existing efforts to mainstream risk reduction are often not sustainable (which is, amongst other things, related to a misinterpretation of the concept of "mainstreaming" that is commonly understood as an "adding-on process" of specific risk reduction activities.) However, genuine mainstreaming signifies the modification of a specific type of core work in order to take into account a new aspect, such as risk reduction and

act upon it. In addition, mainstreaming is generally only considered at the local programme level, rather than at the other institutional levels of the implementing donor or educational bodies etc. Such bodies are needed to ensure the sustainable mainstreaming of disaster reduction.

(c) *There is often a considerable gap between what households and organisations undertake to deal with risk (impeding sustainable risk reduction). The huge variety of existing household strategies and the related financial efforts to cope with risk and disasters show that the vicious circle of increasing risk that can result is not a passive behaviour circle but is related to a combination of weak coping and inadequate outside assistance. This is not on account of any lack of local and institutional efforts; rather it is about their failure to work together.*

As you know, these lessons led to the development of the operational framework for integrating risk reduction, which combines seven complementary strategies for the integration of disaster risk reduction (into recovery programmes) with five complementary measures for reducing risk.[2]*'*

In addition, two (more personal/philosophical) lessons which (also) apply to housing/settlement recovery:

1. *'The devil is in the detail' and, thus, general recommendations are mostly of little help in practice.*
2. *'"Always focus on the positive, not the negative!" And, thus, it is crucial to keep the focus on existing capacities – both local and institutional.'*

Christine Warmsler
Architect/Planner, Researcher/Lecturer/Writer, Lund University

2.7. The challenge to create safe dwellings

'The issue can be seen in the workshop title: "Shelter after Disaster".'

'We should address safer shelter before the disaster and experience shows that it can be done. We need to do much more where we know what the risks are. We have to promote action to address risks so that the hazard does not become a disaster. Disasters create a short-term environment encouraging action about risk reduction, but even then conflicting emergency priorities dampen enthusiasm for longer-term safety and the investment required.'

'Our Development Workshop experience is that once shown how safety can be integrated into existing and new non-engineered buildings families are more than ready to invest their time, energy and money in making their home safer and better for their future. We have to back and encourage this. This is not an issue of building regulations, but of awareness raising on the potential for safety and much lower costs than recovery.'

John Norton
Architect/Planner, Programme Manager, Co-Founder and Executive Director of 'Development Workshop', Recipient: UN Sasakawa Certificate of Distinction 2009 Lauzerte, France

2.8. Why are we rebuilding vulnerability?

'I realise, though looking at lots of earthquake damage, that the actual rebuilding rarely takes adequately into account the need to protect against the next event.'

'When we first looked at reconstructed Pattan and Quetta in Pakistan in 1980, it was clear that (whatever good codes or guidelines had been developed) rebuilding had largely reproduced pre-existing forms of building. My best understanding of the rebuilding after the Kashmir earthquake is that much of it is just the same as before. Has the rebuilding in Turkey after 1999 led to better buildings – at the moment we can't know, but I suspect largely not. And in Haiti, there it seems there is evidence that the rebuilding being done so far is just the same as before. And if this is the case, the larger problem – how to raise standards in the adjacent areas which are more likely to be affected by the next event – is even further from being addressed.'

Robin Spence
Civil Engineer, Specialist in Earthquake Engineering, Researcher, Consultant, Lecturer – University of Cambridge, Founder and Director of 'Cambridge Architectural Research' (CAR) Cambridge, UK (Coburn and Spence, 2002)

2.9. Investment in building knowledge, skills and capacities will create a lasting legacy

'Shelter is essential for survival, and a catalyst for recovery but often also a culprit in causing disasters. Earthquakes do not kill people, unsafe buildings do. Communities living in poor quality shelter on marginal land with inadequate drainage are more vulnerable to floods and landslides; and a few more roof nails and some bracing is all that may be need to ensure simple houses survive a hurricane. Without decent secure shelter families are unable to move beyond survival and build resilient communities that are able to deal with shocks and stresses. Yet, post-disaster the role of shelter in reducing (or increasing) vulnerability long term seems to get forgotten. An important lesson is that investment in building knowledge, skills and capacities to build safe communities based on local technologies and materials is a cost effective way to provide shelter assistance and will leave a legacy that is much more than bricks and mortar.'

Jo da Silva

Engineer, Director – Arup International Development; Lead author of 'Lessons from Aceh'; contributing author to 'Build Back Better'; independent consultant to UNHCR and various NGOs (da Silva, 2010)

2.10. Oysters create their own shells...

My first message learned from Suleiman, carpenter in Akshalaan village near Gediz, Western Turkey, April 1970:

'See those houses that are still standing? They are mine. There are 274 joints in these houses. Most people put one nail in each joint. I put two. That's the difference: 274 nails. At the time the nails seemed expensive.'

My second message learned from Giuseppe Grandori:

'We engineers and architects seem to believe that a community is like an oyster. If we build the shell, we expect the soft living material will appear. In fact, it is the soft human community that must survive and secrete its own shell.'

Frederick Krimgold

Architect/Planner/Director Disaster Risk Reduction Program (DRR) Virginia Tech Advanced Research Institute (ARI) (Krimgold, 1974)

2.11. Organic shelter

'I have learned many lessons from the rebuilding and reconstruction processes of shelters in Gujarat earthquake 2001. The single most important lessons among others is that shelter must grow (with participation of communities) and not simply be built (by contractors). Shelter is organic, it evolves, with time and resources and needs, especially for the poor. On the tsunami experience, I have learnt that in spite of having enough, many people have to stay much longer in temporary houses. Transition shelter or temporary shelter can become the need of outsiders more than the victims. The reconstruction takes time and needs lot of advocacy on inclusion and proper settlement planning. It is recognized that shelter building must take into account the specific needs of the victims.'

'With our Cash for Shelter programme in Kashmir after the earthquake, at least three positive results were learnt:

Firstly, the people built their own houses when given money and time.

Secondly, the money they received for finishing their houses enabled them to buy things they needed, from food to other household items or in other words building shelter was an economic activity generating income.

Thirdly, in this way the local economy was also encouraged and furthermore the people were not dependent on continued external assistance.'

Mihir Bhatt

Architect/Planner, Founder and Director, All India Disaster Mitigation Institute (AIDMI) Ahmedabad, India

2.12. How can we learn from local building practice and avoid intermediate/transition housing?

'Learn from the local people and their building practices. That, for me, is the most important lesson that I have learnt from my experiences. I have found that instant mobilisation towards permanent shelter, with

participation of affected families from design to implementation stages, using local materials, and based on traditional technologies, is the best way forward in a post disaster situation. Intermediate shelters, though gaining popularity with aid agencies in recent years, serve the purpose of sustaining vulnerability of the affected families and the businesses of the aid related industry and nothing else, and must be avoided.'

'The value addition that can, and must, be brought in by practitioners such as us, is of introducing low cost risk reduction technologies within the traditional houses, and any additional environmental features that can help cope with climate variability related risks that are new to the areas. Regarding climate change mitigation, again traditional practices usually offer the most viable solutions and the smallest footprints.'

Anshu Sharma

Architect/Planner, Co-Founder and Director of SEEDS, Delhi India

2.13. Shelter and recovery: The role of transition housing

'Shelter after disaster has extremely important cultural and community aspects. Both rebuilt housing and transitional shelter must seek to preserve community and ensure that it functions in the aftermath of disaster. This includes social support networks and economic and logistical aspects of daily life, as well as the spiritual and psychological aspects of being in a community. As far as possible, shelter should not stigmatise or punish people for being disaster victims. It should facilitate their recovery by allowing them to conduct or rebuild their normal economic activities and to engage in normal kinds of social interaction. Social support and solidarity should not be inhibited by the choices made in the provision of shelter, which should seek not to make divisions in society but to heal them. Shelter should not stigmatise or "ghettoise" people as disaster victims, and neither should it place them in situations of renewed vulnerability.'

'Shelter that is poorly designed from the functional, community and environmental points of view reduces the opportunities to recover from disaster. Well-designed shelter that is made with sensitivity to these issues can facilitate recovery, especially if it encourages social cohesion and the sort of participatory processes inherent in good governance. Shelter is not merely about housing, but also about accommodating other activities, from workshops to worship, to community gatherings, clinics, libraries, and so on. Absence of attention to these needs, and excessive concentration on providing only housing, will restrict the opportunities for recovery from disaster.'

'Unless vernacular housing in a particular area can be reconstructed with few materials and little investment of time and money, transitional shelter enters into the equation. There are unresolved issues about how "permanent" transitional housing should be. More needs to be known about the balance between investing in transitional shelter and permanent reconstruction. With the usual shortage of funds, concentrating on one of these issues would tend to restrict the other. In an ideal situation, transitional housing will be designed to provide sufficient support, and to last long enough, to tide people over until a vigorous programme of permanent reconstruction has been completed. This also requires thought to be given to how to achieve the transition from transitional accommodation to permanently rebuilt environments. With all of these questions there is no single, clear-cut answer, but there is also currently a shortage of knowledge and expertise.'

David Alexander

Geographer, Co-Editor: Disasters Journal, Chief Senior Scientist, Global Risk Forum (GRF) Davos

2.14. Why do we fail to learn lessons from local knowledge and past experiences?

'Every disaster is a new and exciting learning opportunity for me, as I hope to do it better each time. The work process I go through, as a shelter practitioner is almost the same. I have learnt to learn from the in-country staff and local people on what they know and find out what they need to deal with a disaster of a massive scale.'

'I feel very sad to see that we in the humanitarian relief system do not learn how to do it better, as we fail to capture the lessons learnt and move forward. It feels as if we are going backwards by delivering shelter as a relief product and standardising it, in the name of "humanitarian intervention". I wish we had more time to review in advance, before jumping into any new disasters concerning ways to make the shelter interventions that are a more durable solution for future risk reduction and not

repeat the mistakes we are likely to make when we are in a hurry.'

Rumana Kabir
Architect/Independent Shelter and Housing Consultant in Bracknell UK

2.15. Spiders webs and acupuncture...

'A disaster is just another chapter in the community history, and that no matter how deep its impact is, its real memory will go no further than one generation. As outsiders, our challenge (and the most we can do) is to contribute to reinforce local actors, so they can push the community future according to their own priorities and worldviews. We must act as acupuncturists, putting our "needles" (knowledge, experience, money, human support, etc.) into "points" of the existing processes were they can help to restore communication and energy flows between human actors, between the community and the government, between development and ecosystems, between the past, the present and the future.'

'We cannot "direct" those processes; much less determine their objectives or impose our priorities and views. We cannot reconstruct destroyed territories, or avoid disasters to happen. Our duty is not to rebuild the affected web, but to enforce the spider identity, wisdom and power, so she can thread stronger (more resistant and resilient) webs by herself.'

Gustavo Wilches-Chaux
Independent Consultant, Researcher, Lawyer, Philosopher, Writer, Community Activist, Reconstruction Manager. Bogota, Colombia

2.16. Shelter shopping lists and acupuncture...

'...People build and think fundamentally the same way everywhere in the world about shelter/house/home, whatever they may call it, and need pretty much the same basic inputs whether building in Oxford or Aceh. I often think what would I do or want if I were in their shoes and find it amazingly the same:

- *always a piece of land and access to cash/credit/or a loan is the key;*
- *sometime, you need something that you can not find immediately in your own environment so try mail order/or travel to get it/or find an alternative for it/or ask an agency for it;*
- *provide your own workforce with your family;*
- *or if you can afford them hire local architects – (Zaha Hadid?!) or builders;*
- *and there is a need for access to information and knowledge on safety...*

The list can go on...

Ideally we all want to build our dream house, or if not, something that closely matches it and finally we settle down for something we can afford, readily available or given to us with the hope to make it match our dream house incrementally. The role of the external is to recognise what the beneficiaries want, have and can do, and provide the missing or the weakest link in this chain that can sometimes be cash, sometimes, materials, sometimes information and knowledge, sometimes a full house, or a combination of these.'

'Acupuncture may often be more useful than a full operation to provide the missing link As for safety, it is always easier to achieve it at one time during reconstruction or through training etc but, hard to sustain over time unless there is strong enforcement, or a sustainable understanding of what safety is by the users. Needs change, priorities change, materials change so does safety. Acupuncture does not work in this case and needs continuous rehabilitation.'

Yasemin Aysan
Architect, Former IFRC Under Secretary for Humanitarian Affaires, Independent Consultant, Member of 'USG Emergency Response and Recovery'. Geneva, Switzerland (Aysan, 1987, 1992, 1995)

2.17. Shelter: Fulfilling aspirations and building confidence

'For most, their home is a manifestation of their lifelong struggles and aspirations, shelter after disaster should, therefore be able to restore their confidence of living safely in the same area and aligned to their original aspirations. Ultimately, it is not just about square-foot area, material or processes used, rather an effort to maintain continuum that would lead to revitalization.'

Manu Gupta
Architect/Planner, Director and Co-Founder, SEEDS Delhi, India

2.18. Addressing the limitations in current policies

'Post-disaster settlement and reconstruction are challenged currently by both the history and structure of the supporting stakeholders, including the humanitarian and development communities, donors, international finance institutions (IFI's) and governments:

- *coherent operational coordination between these stakeholders rarely results in a single strategy or policy that maintains a record of a common approach and decisions, bridging support to displaced and non-displaced throughout the period of response;*
- *within any strategy, the technical planning of activities rarely reaches the level of credible detail expected in commercial construction, integrating rural or urban planning, communal services such as clinics, sector services such as WASH, and cross-cutting topics such as gender and the environment;*
- *response options appear centred around the needs of rural owner-occupiers of houses, which are often not the most vulnerable or the most numerous as, for example, the majority of those affected in urban environments are tenants, often living in multi-story apartments;*
- *reconstruction often takes many years, in order to resolve land rights and other foundations to sustainable recovery, yet there are insufficient options to support accommodation over this process, other than through incremental transitional shelter;*
- *there is no basic common understanding maintained of the total resources available for a response, neither financial resources from remittances to the World Bank, nor material resources from the local environment to regional markets; and*
- *assessment, monitoring and evaluation are based generally on the numbers of families supported with direct assistance, which does not necessarily reflect the priorities of individuals and communities affected which may, for example, prefer.*

In order to meet these and other challenges to supporting the sector, these stakeholders may wish to consider:

- *establishing departments and engaging specialists representing the full scope of activities undertaken, for example there are very few operational land rights specialists or contract managers on staff, able to improve sector response strategically, especially when considering around 50% of funding for recent major responses such as to the Indian Ocean tsunami and Haiti earthquake was spent on shelter, settlement and reconstruction;*
- *developing common technical training, as the first step in professionalising the sector, understanding that approximately 85% of stakeholder staff regionally have no technical training whatsoever;*
- *developing a common technical library and community of practice support comprising guidance, standards, good practice, case studies, training and events so that stakeholders may share the same memory, understanding that approximately 80% of sector knowledge generated operationally and regionally is currently not accessible, yet stakeholders appear enthusiastic to share much of it; and*
- *better engaging the private sector, assuming that involving national and regional construction industries would be helpful to reconstruction activities, as well as to DRR and sustainable recovery.'*

Tom Corsellis

Architect/Planner, Co-Founder and Executive Director: Sheltercentre (Corsellis and Vitale 2005, Sheltercentre 2010)

3. Common themes

This collection of expert opinion reveals many common areas of concern:

3.1. Constraints on shelter provision

- 'funding varies wildly between disasters.... and is often the most critical determinant' (Section 2.1).
- 'the underlying cultural, political, economic and geographic realities govern what is possible – not the capabilities, resources and preconceptions of outside agencies' (Section 2.2).
- 'The scale of post-disaster shelter need that is increasingly emerging is beyond the response capacity of institutional humanitarians, be they governmental or non-governmental' (Section 2.3).
- 'Post-disaster settlement and reconstruction are challenged currently by both the history

and structure of the supporting stakeholders' (Section 2.18).

3.2. The primary role of disaster survivors in providing their own shelter

- 'the primary actors from disaster to recovery are usually the affectees themselves.' (Section 2.1).
- 'a community is like an oyster. If we build the shell, we expect the soft living material will appear. In fact, it is the soft human community that must survive and secrete its own shell.' (Section 2.10).
- 'shelter must grow (with participation of communities) and not simple be built (by contractors). Shelter is organic, it evolves....' (Section 2.11).

3.3. Roles and attitudes needed in assisting groups

- 'external aid providers need to maintain a humble posture as servant and to furnish their assistance as a vehicle to support the local needs as defined by them, not as a "deliverable"' (Section 2.2).
- 'the biggest lesson is the blindingly obvious one of involving people in decisions that affect them' (Section 2.4).
- 'learn from the local people and their building practices' (Section 2.12).
- 'I have learnt to learn from the in-country staff and local people on what they know and find out what they need to deal with a disaster of a massive scale' (Section 2.14).
- 'as outsiders, our challenge (and the most we can do) is to contribute to reinforce local actors, so they can push the community future according to their own priorities and world views' (Section 2.15).
- 'The role of the external is to recognise what the beneficiaries want, have and can do, and provide the missing or the weakest link in this chain that can sometimes be cash, sometimes, materials, sometimes information and knowledge, sometimes a full house, or a combination of these' (Section 2.16).
- 'developing a common technical library and community of practice support' (Section 2.18).

3.4. The functions of shelter

- 'shelter is the foundation block for people's recovery' (Section 2.5).
- 'shelter is not merely about housing, but also about accommodating other activities, from workshops to worship, to community gatherings, clinics, libraries, and so on. Absence of attention to these needs, and excessive concentration on providing *only* housing, will restrict the opportunities for recovery from disaster' (Section 2.13).
- 'home is a manifestation of their lifelong struggles and aspirations, shelter ...should restore their confidence of living safely' (Section 2.17).
- 'people need pretty much the same basic inputs whether building in Oxford or Aceh:
  - ☐ always a piece of land and access to cash/ credit/ or a loan is the key;
  - ☐ you need something that you can not find immediately in your own environment;
  - ☐ provide your own workforce with your family;
  - ☐ if you can afford them, hire local architects or builders;
  - ☐ and there is a need for access to information and knowledge on safety...' (Section 2.16).

3.5. Managing risk

- 'there is often a considerable gap between what households and organisations undertake to deal with risk' (Section 2.6).
- 'disasters create a short term environment encouraging action about risk reduction' (Section 2.7).

- 'the larger problem – how to raise standards in the adjacent areas which are more likely to be affected by the next earthquake event – is even further from being addressed' (Section 2.8).
- 'the role of shelter in reducing (or increasing) vulnerability long term seems to get forgotten' (Section 2.9).
- 'See those houses that are still standing? They are mine. There are 274 joints in these houses. Most people put one nail in each joint. I put two. That's the difference: 274 nails. At the time the nails seemed expensive'. (Section 2.10).
- 'regarding climate change mitigation, traditional practices usually offer the most viable solutions and the smallest footprints' (Section 2.12).

4. Essential approaches to shelter

The following summary of approaches emerges from personal experiences that span almost 40 years. In a brief paper it is not possible to say (or even remember) which specific experiences defined each approach. These were written before seeking the opinions of experts listed above, but there are many similarities of emphasis.

4.1. First essential approach: Empathise with beneficiaries (see also Conclusions Item 5)

> *'...the survivors lacked nothing: they have medicine. They have hot food. They have shelter for the night. Of course, their current accommodation is a bit temporary, but they should view it like a camping weekend.'*
>
> Prime Minister of Italy, Silvio Berlusconi, L'Aquila, 8th April 2009

This quotation must win a prize as a demonstration of acute insensitivity by a national leader, responding in a crass manner to families who had lost relatives, possessions and their homes. The opposite approach is to 'empathise' with disaster survivors, since any official seeking to provide sensitive and effective assistance must seek to place themselves in the mind-frame of the beneficiaries.

4.2. Second essential approach: Consider process, not just products

> *'The plan is not important, it's planning that matters...'*
>
> Dwight D. Eisenhower

Adopting a 'process perspective' is a particular challenge for shelter manufacturers and professionals such as architects or engineers, who tend to regard a tangible product as the natural solution to a given accommodation problem. However, – *'a journey can be just as important as arriving at the destination'*, or in another saying: *'it's the making that matters, not the thing I make'*.

Terminology is important in placing emphasis on process. For example: *'sheltering'*, is more expressive than just *'shelter'*, and *'housing'*, is a more accurate word than just *'houses'* when considering the sheltering process as the first stage in recovery and reconstruction.

In writing the UN Shelter after Disaster Guidelines in the late 1970s the authors, Fred Cuny, Fred Krimgold and myself, with support from Paul Thompson and Aloysius Fernandez, conducted some research on a number of disasters to identify the varied functions of shelter. While some were clearly physical products, all were in essence processes:

1. Protection from sun, cold, wind and rain
2. Storage of belongings
3. Protection of property
4. Establishment of territorial claims (ownership and occupancy rights)
5. The establishment of a staging point for future action (including salvage and reconstruction, as well as social reorganisation)
6. Emotional security and the need for privacy
7. An address for the receipt of services (medical aid / food distribution etc)

8. Shelter within commuting distance from employment
9. Accommodation for families who have temporarily evacuated their homes for fear of subsequent damage

(UNDRO, 1982)

It is rather perplexing that we did not discover a further essential shelter function: *providing a work place for families who use their homes for their daily occupations*.

4.3. Third essential approach: Build capacity

In the 1982 UN Guidelines on 'Shelter after Disaster' we attempted (totally in vain I might add...) to shift the attitudes of relief providers of shelter products by using the term 'survivors' rather than 'victims' since we believed that the expression 'victim' automatically, and falsely, cast them in a dependency category of some helpless persons in need of support from an assisting group.

'Capacity Building' figures in most agency fund raising literature and within most evaluation reports. But the reality in too many cases is that the agencies give priority to building their *own* capacity.

Capacity building is often confined to developing training courses; however, it can cover many additional aspects.

The first area is training and education: Many groups may need to be trained, from builders to professionals to the users of buildings. Training may be to achieve good building practice as well as to ensure hazard safety.

The second area is by building (or rebuilding) institutions: Local institutions play such key roles in recovery that they normally need to be strengthened (or repaired) to fulfil their vital functions. In addition local government may need support as the natural focal point for recovery.

The third area is by social mobilization: Many examples can be cited – such as strengthening the role of host families as they accommodate disaster survivors, or providing a constant stream of information from assisting groups to help survivors in making the choices they face in their recovery.

The fourth area is by strengthening livelihoods: Getting survivors back to work assists them to fund their own recovery and restore some level of normality after the trauma they have suffered.

Thus, the third approach is to give priority to building the capacity of survivors at all times, rather than concentrating on the capacity of the government and other assisting groups or donors.

4.4. Fourth essential approach: Place shelter within the wider context

This integrated approach places shelter and the reconstruction of dwellings in its economic, social and environmental context rather than confining attention to shelter in isolation. If an external contractor builds, then you have a safe shelter or house, as well as associated infrastructure. But if the community participate or create their shelters or dwellings, then the added value in other sectors can be astonishingly impressive.

A well-integrated project will provide valuable psycho-social therapy for the community and in the process provide community solidarity from working together. A locally based shelter and housing reconstruction programme can help strengthen the local economy and impart new skills in building, community organization, project management and financial monitoring.

But the process can yield even further benefits such as strengthening livelihoods, (as noted above in the Third essential approach: building capacity), improve safety and contribute to sustainable environments by close attention to land-use planning and the selection of building materials.

Thus, shelter following disaster cannot be divorced from its context; it is isolated 'at its peril'.

4.5. Fifth essential approach: Respond to 'demand', rather than 'supply' pressures

Survivors have many requirements including the need for shelter and other basic requirements, while assisting individuals, agencies, governmental departments and manufacturers also possess pressing needs to satisfy. And these interests may be in sharp contrast to the needs of survivors.

Thus, the fifth essential approach is to adopt a 'demand driven response' to shelter and reconstruction and resist all pressures from governments, NGO's, consultants manufacturers and suppliers to follow a 'supply driven response' (see Table 1 on page 208. Please note, the pair of demands and supply pressures set out in this table do no always connect).

4.6 Sixth essential approach: Regard shelters and dwellings as 'homes', not merely as houses

> *'A town is made of buildings, but a community is made of people; a house is a structure but a home is much more. The distinctions are not trivial, nor are they sentimental or romantic: they are fundamental to the understanding of the difference between the provision of shelter which serves to protect and the creation of domestic environments that express the deep structures of society'* (Oliver, 1981).

The task for designers (whether professional or indigenous) is to seek to create *'places'* with meaning, not merely spaces. Places that provide *'identity'* and a deep sense of *'belonging'* and *'security'* are essential, and not merely for protection from the elements, or as occupation and accommodation.

Fulfilling these deeper social, emotional and psychological needs of people constitutes a basic priority that is particularly important in disaster recovery situations where disaster survivors need additional security from their physical environment, following the turmoil of human and material losses. A secure shelter or reconstructed dwelling can play a decisive role in the well-being and recovery of a displaced family and community (Oliver, 1981; Aysan and Oliver, 1987).

4.7. Seventh essential approach: The safety of people, property and buildings has to be a paramount concern in reconstruction

> *'Today's "Act of God" will be regarded as tomorrows "Act of Criminal Negligence"'* (Professor N.N. Ambraseys, 1976).

To achieve this goal of a *'safety culture'* it is necessary to:

Understand the threats: By undertaking pre-disaster risk assessment to grasp the threats of all risks, health, fire, theft, violence and relevant hazards.

Teach and inform: By communicating the principles and practice of safe design and behaviour to builders, craftsperson's, building occupants and professional designers.

Plan in advance: By developing relevant and up-to-date disaster preparedness and recovery plans at all levels from the local to the centre.

Legislate: With effective, relevant and well-enforced land-use planning controls and building regulations for safe design and construction.

4.8. Eighth essential approach: Avoid waste and social upheaval by cutting out transitional shelters

Whenever disasters coincide with elections a severe distortion in aid can happen and an extreme example occurred in L'Aquila, Italy following the 2009 Earthquake. With mounting pressure from the Italian press on the government to provide shelter before the anticipated severe winter and with impending provincial and municipal elections, Berlusconi's government provided seven-star provision for the homeless in what must be the most expensive 'Transition Housing' ever provided after a disaster. In one example, 19 sites were built to accommodate about 15,500 survivors in 4,300 apartments. The average cost of each small – 40–60 m^2 apartment, was 280,607 Euros (£259,000). This high cost was mainly due to the bizarre decision to set each

TABLE 1

'Demand' pressures of survivors	'Supply' pressures of assisting groups
▪ To find shelter as near as possible to their previous home ▪ Whenever possible to avoid occupying mass occupancy campsites, especially where cooking and sanitation is communally provided ▪ To move in temporarily with friends and families ▪ To receive cash grants with no strings attached ▪ To use their own skills and labour in organising shelter and reconstruction of their homes (this particularly applies in rural areas) ▪ To receive at no cost or low cost, basic building materials and tools (plastic sheeting, roofing materials, rope, tools, etc.) ▪ To repair a damaged dwelling	▪ For Governments to rapidly provide highly visible shelters and housing, especially at times of elections ▪ For Governments, military and some international agencies to create mass occupancy campsites ▪ For NGO's to spend their allocated funds rapidly, with photographs for use in fund raising ▪ For NGO's to satisfy the expectations of their donors or constituencies ▪ For consultants to secure commissions to plan or design buildings and settlements ▪ For manufacturers and contractors to secure work

temporary housing block on 40 seismic base isolators. This is a sophisticated anti-seismic system of 'cushions' that reduce earthquake shaking, but each unit costs the Italian taxpayers 1,427 Euros (£1,320.00) (Alexander, 2011).

L'Aquila was an exceptional case. Fortunately most prudent governments avoid wasting precious cash on exorbitantly expensive temporary housing that reduces reconstruction funds.

The need for Transition Housing Issue has become a major debate within the Shelter Community with two competing strategies often being advocated:

Strategy 1

Develop a seamless approach that moves from immediate sheltering to a safe permanent reconstructed dwelling.

Advantages:

1. It avoids a double reconstruction effort.
2. It saves cash reserves.
3. It avoids social disruption.
4. It encourages the distribution of building elements to affected families (such as timber, roofing materials, building tools). Such materials can be used in improvised shelters and then reused in the permanent dwelling.
5. It encourages the wider use of host families to accommodate survivors, as well as to provide them with support.
6. It encourages rapid reconstruction.
7. It encourages pre-planning for reconstruction, particularly with regard to safety regulations.
8. It encourages authorities to look for creative ways to provide innovative forms of shelter, such as:
   - use of existing buildings (such as the use of hotels out of tourist seasons).
   - temporary use of unoccupied dwellings.
   - use of mobile homes/caravans/adapted containers/construction workers huts.

Disadvantage:

1. It may not be possible to adopt this strategy where there are harsh climate conditions.

Strategy 2

Develop a three-stage approach that proceeds from immediate shelter, to a transition house, to a safe permanent reconstructed dwelling.

Advantages:

1. It may be necessary where the start of reconstruction is severely delayed.
2. It may be essential in urban areas of industrialized countries where survivor expectations are very high.
3. It may be essential where impending harsh climate conditions are anticipated and it is essential to replace tents and improvised structures with more substantial shelter.
4. It meets the demand from survivors and officials to provide something superior to tents, while awaiting permanent reconstruction.
5. It provides the time for authorities to plan reconstruction properly, ensuring full participation in decision making and in developing building and land-use safety standards.

Disadvantages:

1. It wastes precious financial resources, needed for permanent reconstruction.
2. Often transition houses arrive after permanent dwellings are rebuilt.
3. It can amount to double reconstruction.
4. Transition houses that are not demolished can become instant slums.
5. In congested urban areas the site taken up by a transition house may be the exact site needed for reconstruction, thus making reconstruction impossible.

(Ashmore, 2009a and 2009b).

4.9 Ninth essential approach: Regard shelter and dwelling reconstruction as a development rather than relief/welfare issue

> *'The primary resource in the provision of post-disaster shelter is the grass-roots motivation of survivors, their friends and families. Assisting groups can help, but they must avoid duplicating anything best undertaken by survivors themselves.'*
>
> First Principle, UN Shelter After Disaster Guidelines (UNDRO, 1982) (see Table 2, page 210).

4.10. Tenth essential approach: Build evidence based knowledge since accurate evidence is often non-existent

Project evaluation of shelter following the Gujarat Earthquake:

> *'In the view of our shelter consultant, the results in terms of new settlements have been poor, particularly in responsiveness to individual needs, and recognition of the essential functions of housing in relation to livelihoods. The process involves a skewing of DEC resources towards the tiny numbers of beneficiaries of these projects. It seems likely that many of these houses will not be used by the owners and may end up as 'second homes'. This suggests a serious problem in relation to organisational learning'* (Disasters Emergency Committee, 2001).

Despite almost 40 years of UN, NGO and governmental experience of shelter needs and provision there has still been minimal research on a wide range of key issues (Davis, 2009).

Social impact:

- role and limits of host families providing accommodation,
- effect of relocations of communities,
- preferences of survivors for varied forms of shelter,
- precise functions of shelter,
- social and cultural aspects of shelter/reconstruction.

Settlement planning:

- long-term impact of shelter and transition housing on human settlements,
- specific problems of shelter needs with dense urban environments,
- the growth of safety in new buildings and their siting,
- the role and performance of transition houses.

Shelter policies:

- overall effectiveness of the three overlapping UN-sponsored shelter clusters.

TABLE 2

Characteristics of a welfare approach to shelter provision	Characteristics of a developmental approach to shelter provision
▪ Projects conceived by relief agencies who *direct* programmes ▪ External assessments are made of shelter needs ▪ Beneficiaries are regarded as 'passive victims' needing support ▪ Absence of cash grants and trust by assisting groups ▪ The overall stance is one of hand-outs of free shelters and shelter materials ▪ Minimal participation of survivors in the shelter or housing reconstruction process (planning and building)	▪ Projects conceived by development agencies who *facilitate* programmes ▪ Where possible survivors assess their own shelter needs ▪ Beneficiaries are regarded as 'active survivors' in organizing their own sheltering or reconstruction ▪ Cash grants are provided to enable survivors with resources to purchase their own shelter materials, (or other essentials) demonstrating trust by assisting groups ▪ Where possible survivors pay in cash or kind for their own shelter ▪ Full participation of survivors in the shelter or housing reconstruction process ▪ Shelter is closely linked to livelihood development and other social and environmental sectors ▪ Training opportunities are grasped at all levels and for any activity that supports recovery

Source: (Abbas et al., 2009; Baraket, 2003).

5. Conclusions

The *'Summary of Progress from 1972–2011'* set out in Section 1 indicated the massive growth in agency interest in post-disaster shelter within the past 5 years as well as in the positive development of safety and user-build housing reconstruction.

The 'Expert opinions' and 'Common themes' sections provided strong reinforcement of views concerning the severe funding, cultural and economic constraints on shelter provision by external groups. They also emphasized the primary role of disaster survivors in providing their own shelter and the need for participation of beneficiaries at every stage in their recovery. The diverse functions of shelter were also highlighted as well as the paramount importance of safe reconstruction.

Finally, I have attempted to summarize my own learning experiences in Section 4: 'Essential approaches' and there is a high degree of consensus with the insights from experts in the field contained in the previous sections.

These essential approaches all flow from an attitude, policy, strategy or practice that is based on the words and example of Christ:

> *'...in everything, do to others what you would have them do to you'* (Matthew 7: 12).

The simplicity of this familiar saying can easily conceal its implicit demands. It throws down a challenge to our all pervasive self-interest and defines an attitude of respect that can form the bed-rock of all shelter and reconstruction policies and practices, shape all recovery actions and affect all dealings with the affected population and officials, as well as with other assisting groups (Tearfund, 2011).

Notes

1. The Humanitarian Review, Ashdown (2011) discussed the cluster approach to shelter (p. 25). 'The IFRC was nominated as the cluster lead agency, and this has not worked well. Partly this is because shelter is difficult. Partly it is because there is an

artificial split whereby IFRC is responsible in the immediate aftermath of a disaster, but not beyond the emergency phase of displacement'.

2. I am referring to seven strategies for integrating risk reduction into recovery Planning: (1) implementing direct stand-alone actions, (2) implementing direct integrated risk reduction, (3) programmatic mainstreaming of risk reduction, (4) organizational mainstreaming of risk reduction, (5) internal mainstreaming of risk reduction, (6) the creation of synergies for risk reduction and (7) educational mainstreaming of risk reduction into recovery. In addition there are five ways to reduce risk include: (1) hazard reduction, (2) mitigation, (3) preparedness, (4) risk 'financing' and (5) preparedness for recovery. These lie at the heart of the operational framework for the integration of risk reduction into disaster recovery.

References

Abbas, K., Jah, A. K., Barenstein, J. D., Phelps, P. M., Pettit, D. and Sena, S., 2009. *Safer Homes, Stronger Communities. A Handbook for Reconstruction*. GFDRR, The World Bank, Washington, DC.

Alexander, D., 2011. Private communication with author. May 2011

Aristide, J. B., 2011. *The Guardian*, 5 February 2011, 39.

Ashdown, P., 2011. *Humanitarian Emergency Response Review*. Senior Advisory Board chaired by Lord (Paddy) Ashdown DFID, London, 28 March 2011. www.dfid.gov.uk/emergency-response-review

Ashmore, J., 2009a. *Shelter Projects, 2008*. UN HABITAT and IFRC, Geneva.

Ashmore, J., 2009b. *Shelter Projects, 2009*. UN HABITAT and IFRC, Geneva.

Aysan, Y. and Oliver, P., 1987. *Housing and Culture after Earthquakes, A Guide For Policy Making on Housing in Seismic Areas*. Oxford Polytechnic, Oxford.

Aysan, Y. and Davis, I. (eds), 1992. *Disasters and the Small Dwelling, Perspectives for the UN IDNDR*. James and James, London.

Aysan, Y., Clayton, A., Cory, A., Davis, I. and Sanderson, D., 1995. *Developing Building for Safety Programmes, Guidelines for Organizing Safe Building Improvement Programmes in Disaster-Prone Areas*. Intermediate Technology Publications, London.

Baraket, S., 2003. *Housing Reconstruction after Conflict and Disaster*. HPN Network Paper 043, Overseas Development Institute (ODI), London.

Clayton, A. and Davis, I., 1994. *Building for Safety Compendium, an Annotated Bibliography and Information Directory for Safe Building*. Intermediate Technology Publications, London.

Coburn, A., Hughes, R., Pomonis, A. and Spence, R., 1995. *Technical Principles of Building for Safety*. Intermediate Technology Publications, London.

Coburn, A. and Spence, R., 2002. *Earthquake Protection*, 2nd edn. Wiley, Chichester.

Corsellis, T. and Vitale, A., 2005. *Transitional Settlement: Displaced Populations*. Oxfam Publishing, Oxford.

da Silva, J., 2010. *Lessons from Aceh: Key Considerations in Post-Disaster Reconstruction*. Practical Action Publishing, Rugby.

Davis, I., 1978. *Shelter after Disaster*. Oxford Polytechnic Press, Oxford.

Davis, I. (ed.), 1981. *Disasters and the Small Dwelling*. Pergamon Press, Oxford.

Davis, I., 1985. *The Provision of Shelter following Earthquakes*. PhD dissertation, Development Planning Unit (DPU), University College, London (UCL).

Davis, I., 2009. Historical introduction and review of historical shelter projects. *Shelter Projects*, J. Ashmore (ed.)., UN HABITAT and IFRC, Geneva. 85–120.

Dudley, E. and Haaland, A., 1993. *Communicating Building for Safety, Guidelines for Methods of Communicating Technical Information to Local Builders and Householders*. Intermediate Technology Publications, London.

Disasters Emergency Committee, 2001. *Evaluation Report of DEC Assistance following the Gujarat Earthquake*, Disasters Emergency Committee (DEC), London.

Fleuristil, A., 2011. *The Observer*, 1 January 2011, 18.

IFRC, 2010. *Owner-Driven Housing Reconstruction Guidelines*. IFRC, Geneva.

Krimgold, F., 1974. *Doctoral Thesis on Disaster Prevention* (with specific reference to Shelter issues). Royal Institute of Technology, Stockholm.

Lyons, M. and Schilderman, T., (eds) 2010. *Delivering People-Centred Reconstruction at Scale*. Practical Action Publishing, Rugby.

Oliver, P., 1981. The cultural context of shelter provision. *Disasters and the Small Dwelling*, I. Davis, (ed.). Pergamon Press, Oxford. 39–42.

Saunders, G., 2011. Communication with author. May 2011.

Sheltercentre, 2010. *Shelter after Disaster: Strategies for Transitional Settlement and Reconstruction*. Sheltercentre and UN/OCHA. Replacement of UN Shelter after Disaster Guidelines (1982), Geneva.

Sphere, 2004. *The Sphere Project, The Humanitarian Charter and Minimum Standards in Disaster Response*. Sphere and Oxfam Publishing, Geneva.

Tearfund, 2011. *Guidelines on Shelter and Reconstruction*. Tearfund, Teddington.

UNDRO, 1982. *Shelter after Disaster – Guidelines for Assistance*, I. Davis (ed.), with Cuny, F., Krimgold, F., Thompson, P. Office of the United Nations Disaster Relief Coordinator (UNDRO), New York.

UNOCHA, 2011. *Shelter after Disaster – Guidelines for Assistance – Refreshed Version*, I. Davis (ed.), with Krimgold, F., Thompson, P. United Nations Office of Conflict and Humanitarian Affairs, Geneva.

Harnessing time: Reflections on constraints to development

MICHAL LYONS[1], THEO SCHILDERMAN[2] AND DAVID SANDERSON[3]

[1]Social and Policy Studies, Faculty of Arts and Human Sciences, London South Bank University, London, UK
[2]Practical Action, International Policy and Programmes, The Schumacher Centre for Technology and Development, Bourton on Dunsmore, Rugby, CV23 9QZ, UK
[3]Centre for Development and Emergency Practice (CENDEP), School of the Built Environment, Oxford Brookes University, Gipsy Lane, Oxford, OX3 0PB, UK

This short piece argues that a series of progressive developments in our understanding of aid after disasters, and how we should seek to deliver and manage it, calls for a review of accountability systems in the sector.

In 1984, Weijkman titled his influential book on the causes of natural disasters 'Natural Disasters: Acts of God or *acts of man*?' (authors' emphasis). His now widely accepted argument was that disaster is created by vulnerability to natural hazards. Both vulnerability to disaster and the resilience necessary to recover from it are inextricably linked to man-made circumstances. These circumstances operate at different levels, often in interaction with one another. Crudely, one might identify a global level as affecting vulnerability and resilience through war, demographic change, poor governance; a national level, through dynamic pressures such as land tenure, urbanization, economic policies and epidemics; and a local level, through unsafe conditions created by poor land management, poor construction and chronic hunger.

A complex system of politics, economics, culture and demographics links not only the different causes of vulnerability and resilience, but also the different levels of cause. The progress of globalization and the increasingly wide adoption at local, national and international levels of neoliberal economic policies and of legislation and policies designed to facilitate it, and the accompanying growth of world trade, have been associated with growing in-country inequality over the last quarter of the 20th century, only partly moderated by some subsequent falls. These have further widened the gulf between more and less vulnerable fractions of society, reduced mutual trust and reduced popular support for redistributive policies (Toya and Skidmore, 2010; Donner and Rodriguez, 2008; Meschi and Vivarelli, 2009; Wilkinson and Pickett, 2010; Mutter, 2010).

The changing environment has profoundly affected the architecture of aid. The sector has experienced concentration, with an increasing share of funds going to a relatively small number of large organizations; the increasing competition for funds and increasing public suspicion for redistributive policies of any sort have forced an increasing emphasis on 'upward' accountability to donors, increased emphasis on measurable outputs and adherence to planned programmes and projects (e.g. Rogerson, 2005).

This comes at a time when, increasingly, it has become apparent that development and disaster risk reduction are interdependent. For example, income inequality and quality of institutions have been linked to the rate of fatalities from earthquakes. In the wake of Hurricane Katrina, some have argued that natural disasters further increase socioeconomic inequalities resilience to disasters and disaster risk reduction are both intimately linked to development. The resilience of, and the reduction of vulnerability reduction in settlements, cities and countries cannot be achieved by the construction of a technically sound building stock alone, but depend directly on the resilience or vulnerability of their populations. Even sound buildings will degrade over time if not properly maintained, yet maintenance depends on knowledge and means; continued maintenance of buildings' resilience over time requires adaptation to changing distributions and types of hazards brought about by, for example, changing land management and land use. For local populations to adapt to these they need to be aware of the changes, their implications for vulnerability and appropriate building adaptations. Yet poverty and powerlessness are isolating, and therefore unlikely to provide conditions for the acquisition of the necessary information and access to appropriate responses. Thus, particularly for the poor, the development of resilience and the reduction of their vulnerability means marshalling, drawing upon and fostering the complex combination of social, financial, natural, physical and human resources necessary for long-term development, sustainability and poverty reduction (HOC, 2005; Kahn, 2005; Anbarci et al., 2005).

Thus, the humanitarian response to disasters must confront the challenges of supporting development, in order to meet its own goals of supporting recovery, developing resilience and reducing vulnerability. Ultimately, the institution and maintenance of a continuous framework of support for the achievement of complex development goals through disaster response, requires flexibility over time frames and the development of multifaceted programmes. Traditionally, the shelter sector has not been structured to deliver the necessary programmes, being isolated from other sectors and subject to the rigid phasing of disaster response. It has followed the precepts of the Humanitarian Charter, but their agreed interpretation, embodied in the widely respected Sphere project to develop universal minimum standards, begun in 1997, originally accorded shelter a limited role, seeing it as necessary to survival but not necessarily as part of a broader and more ambitious developmental process. In 2004, a slightly broader interpretation was adopted, reflecting growing unease with a stark survivalist approach, but still does not link shelter to processes of livelihood development and sustainability or poverty reduction 'Everyone has the right to adequate housing. This right is recognised in international legal instruments and includes the right to live in security, peace and dignity, and with security of tenure' (Sphere, 2004, p. 207; Van Dyke and Waldman, 2004).

Yet three important trends indicate that, today, the sector is moving beyond these limitations to meet the challenges it faces. While national governments pull back from redistributive policies, the shelter sector, with support from multilateral and bilateral organizations, has begun to adopt an increasingly long-term, developmental and redistributive perspective on its work. Although the process has been underway for some 15 years, it received great impetus following the South Asian tsunami. Partly as a result of the large funds raised for this disaster, agencies that had previously specialized in emergency response expanded their activities into reconstruction; and agencies that had specialized in development entered the post-disaster reconstruction arena. The resultant experiences gained by agencies, and the new partnerships and alliances formed, have been profoundly fruitful for the sector (Lyons, 2010).

The first important trend is the erosion of a previously rigid phasing of response activities. The changing architecture of post-disaster aid, which followed the Asian tsunami, has brought together development and emergency organizations in an unprecedented collaboration, and re-energized an already dynamic and ongoing questioning of the singularity of post-disaster phases and the extent to which they can be regarded as separate. It has also contributed to a growing emphasis on the developmental role of emergency assistance, as is clear, for example, from the interpretation of necessary competencies for humanitarian professional, which emphasize skills and outcomes necessary for development work (Walker et al., 2010); or from the OCHA guidelines on transitional settlement and reconstruction after natural disasters. These emphasize the role of shelter and the approaches taken to its delivery in 'maximizing communal coping strategies' (OCHA, 2008, p. 2), the provision of support regardless of tenure status and the recognition of currently excluded groups as full beneficiaries, the importance of risk reduction and the minimization of exposure to hazards (OCHA, 2008, pp. 3, 4). Most importantly, they emphasize the need to prioritize a continuous, uninterrupted and strategically conceived effort in the phasing of responses and the handover of responsibility from one agency to another. While implementation of these principles has been hampered by a range of political and practical issues, their adoption is to be welcomed. For several years now, the sectoral clusters operating in the field have been instrumental in taking forward the debates on implementation of these principles, and the development of common approaches on a continuum such as emergency shelter-transitional, shelter-permanent reconstruction. The continued work of the Shelter Cluster, both outside and in collaboration with the Shelter Centre, provide a number of useful instances.

Importantly, the policy documents are enshrining best practice, rather than providing top-down advice. For example, again in Sri Lanka, a number of humanitarian NGOs, which originally raised funds only for the relief phase, were able to receive permission from their donors to reallocate their funds to the construction of permanent housing and the achievement of other development goals.

Second, stringent criticism continues to be levelled at the humanitarian sector as a whole over failures of coordination. However, there is a relatively new – and as yet developing drive – to harmonize philosophies, goals and standards of disaster response and to better coordinate work among agencies and with governments. A great deal of effort is being put into this drive by agencies, NGOs and formal and informal consortia. The sectoral cluster structure, which facilitates the breakdown of strict phasing and the development of shared ideas on within-cluster approaches, is less effective as a vehicle for comprehensive programming of multi-sector work.

The third important trend has been the growing adoption of a development agenda within reconstruction programmes. As alluded to earlier, reduction of vulnerability to disasters requires a view of reconstruction which goes beyond the production of safe buildings or provision of non-food items to encompass sustainable livelihood development in its broader sense. This has increasingly been recognized by humanitarian sector policy documents and guidelines, and Noradd, for example, has attempted to capture the necessary complexity by drawing on DFID's Sustainable Livelihoods Framework. Originally developed to describe the ability of the rural poor to survive adverse trends and shocks, this framework has subsequently been used by development scholars and development organizations everywhere to describe the poor's adjustments to urban environments and a wide range of challenges. Its key insight is the wide range of resources marshalled – and needed – for survival, including natural, physical, financial, social and human 'capital' (Scoones, 1998; Rakodi and Lloyd Jones, 2002; Ashdown, 2011; Crawford and Suvantne, 2010).

Again, the tsunami response from early 2005 onwards provided a milestone in this process, demonstrating that people-centred reconstruction programmes have both developmental and practical benefits, which included both lower unit production costs and times, and better support for the re-emergence of livelihood activities and local-political organization. As well as highlighting the challenges of such approaches for development and inclusiveness, an important lesson from this experience has been the viability and practicability of such approaches (Lyons, 2009).

The need to meet the development challenges of post-disaster response and reconstruction, places greater demands than ever on NGOs to be adaptable to conditions on the ground. For example, the achievement of development goals is threatened in many reconstruction programmes by the exclusion from access to support aid of the most vulnerable populations. In Sri Lanka, following the tsunami, many NGOs eventually acquired land on which displaced households could build their homes, enabling them to access the compensation stipulated in the national housing programmes and any top-up funds, which were restricted to those who could prove pre-existing title to land. Similarly, NGOs that had expected simply to build houses and leave, found it necessary to partner with other specialist NGOs to support income-generation or educational programmes. In many post-disaster environments, under pressure to demonstrate rapid results and under the eye of international media, there are tensions between governments and incoming NGOs. These can be exacerbated during the reconstruction process by the reluctance of NGOs to address complex local specificities. For example, the sheer complexity of local land, planning and other related legislation and institutions has been an obstacle in the way of collaboration between agencies and governments, and delayed or prevented the development of some large-scale projects.

Thus, to summarize, a key feature of all three trends in humanitarian sector approaches to reconstruction is growing flexibility: flexibility over programming substance, timing of interventions, adaptation of interventions to realities on the ground and so on. Yet, despite these developing agendas, the drive for change is constrained by upward accountability structures that reduce flexibility on the ground. In an era in which aid is increasingly constrained by donor government foreign interests, resulting in the 'securitization of aid'; and in which – in part due to increasing public suspicion over redistributive policies – accountability to donors has been increasingly required to be precisely articulated in terms of direct and measurable relationships between programme objectives, outcomes and outputs, it remains difficult to be flexible and responsive on the ground. For example, it is sometimes evident after the emergency phase that transitional shelter is not needed in the short term and is unlikely to contribute to long-term reconstruction, yet funding has been allocated on the assumption that it will be used for that purpose. A perverse outcome of this accountability model has occasionally been that NGOs have gone ahead with projects whose usefulness they have come to doubt.

At the same time, there are growing calls in the development sector for 'downward', as well as 'upward' accountability. This is increasingly considered both a moral imperative and, more importantly for our argument, an enabling factor for development outcomes. However, a strong emphasis on upward accountability pre-empts this: a major barrier to downward accountability is presented in the case of post-disaster interventions by the fact that beneficiaries have no say in setting priorities or in critical planning decisions. A second barrier is presented by the fact that, even when project outcomes are reported to beneficiaries, who might wish to contest them, they have only advisory power in the monitoring and evaluation process. Executive and financial powers lie with the NGO and the donors. Lack of satisfactory accountability to donors risks direct penalties for NGOs in terms of, for example, access to future funds. On the other hand, lack of satisfactory accountability to beneficiaries carries only less tangible risks for the organization.

Drawing together these strands, an argument could be made that the programming flexibility which is increasingly demanded by practitioners in the field should be reflected in the institutional structure of the sector. This could be done by the establishment of a financial mechanism for holding funds while their use is being reconsidered. For example, if unanticipated institutional, market or physical obstacles to execution of planned projects become apparent, or if planned projects turn out to be redundant, the funds earmarked for them could be invested in such a fund and their use reconsidered. One possible model would be a trust fund, with trustees representing donors, the NGOs, beneficiaries and relevant government officials. Opening up discussion over fund allocation to beneficiary groups would give them some power in the reconstruction process and therefore tie in well with the growing call for 'downward', as well as 'upward' accountability. Programmes subjected to review by such a panel – and in the light of mature reflection over the circumstances of disaster – are likely to provide a more developmental set of inputs, outputs and outcomes and to facilitate continuity of development and coherence of policy and programming. We have given ourselves leave to review our operations. Now it is time to review our systems.

References

Anbarci, N., Escaleras, M. and Register, C. A., 2005. Earthquake fatalities: the interaction of nature and political economy. *Journal of Public Economics*, 89. 1907–1933.

Ashdown, P., 2011. *Humanitarian Emergency Response Review*. DFID, London. www.dfid.gov.uk/emergency-response-review.

Crawford, K. and Suvantne, M., 2010. *New Urban Assistance Guidelines: Launch, Care International UK and Norwegian Refugee Council*. Shelter Centre, Geneva. http://sheltercentre.org/sites/default/files/SM10a_Day2_0945_New_urban_guidelines_Kate_Crawford_ARE__Martin_Suvatne_NRC.ppt.

Donner, W. and Rodriguez, H., 2008. Population composition, migration and inequality: The influence of demographic changes on disaster risk and vulnerability. *Social Forces*, 97(2). 1089–1114.

House of Commons International Development Committee. 2005. *Humanitarian Response to Disasters*. HOC. 1–1188.

Kahn, M., 2005. The death toll from natural disasters: the tole of income, geography and institutions. *Review of Economics and Statistics*, 87. 271–284.

Lyons, M., 2009. Building back better: on the large-scale impact of small-scale approaches to post-disaster reconstruction. *World Development*, 37(2). 385–398.

Lyons, M., 2010. Development from disaster: evaluating post-disaster housing reconstruction as development. *Building Back Better: Delivering People-centred Reconstruction at Scale*, M. Lyons and T. Schilderman (eds). Chapter 2 (2009) Practical Action Publications, Rugby. 39–61.

Meschi, E. and Vivarelli, M., 2009. Trade and income inequality in developing countries. *World Development*, 37(2). 287–302.

Mutter, J., 2010. Disasters widen the rich-poor gap. *Nature*, 466. 1042.

OCHA. 2008. *Transitional Settelment and Reconstruction after Natural Disasters*, Field edn. United Nations, Geneva.

Rakodi, C. and Llyod-Jones, T., 2002. *Urban Livelihoods; a People-centred Approach to Reducing Poverty*. Earthscan Publications, London.

Rogerson, A., 2005. *The International Aid System 2005–1010, Forces For and Against Change*. www.odi.org.uk/publications/web_papers/aid_system_rogerson.pdf.

Scoones, I., 1998. *Sustainable rural livelihoods: a framework for analysis*. Working Paper No. 72, Institute of Development Studies, University of Sussex.

Sphere Project. 2004. *The Sphere Project: Humanitarian Charter and Minimum Standards in Disaster Response*. Oxfam Publishing, Geneva.

Toya, H. and Skidmore, M., 2010. *Natural Disaster Impacts and Fiscal Decentralisation*. FESifo Forum 2/2010. www.cesifo-group.de/portal/page/portal/DocBase_Content/ZS/ZS-CESifo_Forum/zs-for-2010/zs-for-2010-2/forum2-10-focus5.pdf.

Van Dyke, M. and Waldman, R., 2004. *The Sphere Project Evaluation Report*. Mailman School of Public Health, Columbia University, New York.

Walker, P., Feinstein International Center, Tufts University and Russ, C., 2010. *Professionalising the Humanitarian Sector: A Scoping Study*. ELHRA, April 2010 www.elrha.org/uploads/Professionalising_the_humanitarian_sector.pdf.

Wilkinson, R. and Pickett, K., 2010. *The Spirit Level – Why Equality is Better for Everyone*. Penguin Books, London.

Post-disaster reconstruction: A current analysis of Gujarat's response after the 2001 earthquake

PLATO JACK POWELL

Independent Researcher, Bristol, UK

Post-disaster reconstruction is a complex process that involves the interaction of social, technological and economic factors. The most important goal of any post-disaster reconstruction programme must be to reduce the long-term vulnerability of affected communities through the construction of multi-hazard proof housing and appropriate knowledge transfer. Post-disaster reconstruction is an ever-evolving process and there is by no means a perfect solution that will apply to every disaster. It is essential that academics, governments and humanitarian agencies learn from previous applications of this process in order to not only implement programmes that reduce the current vulnerability but also to establish a legacy of disaster-risk reduction. This paper examines the current structural vulnerabilities of different post-disaster reconstruction approaches implemented after the 2001 Gujarat earthquake. Field research involved visual surveys and questionnaires. This research highlighted a number of issues that would affect the current vulnerability of the buildings' stock in the event of further earthquakes. The paper highlights that there was a high level of technical knowledge transfer in Gujarat after the earthquake, but unfortunately this knowledge transfer did not shape a legacy of disaster-risk reduction. There were also a number of significant issues that were the result of inadequate policies implemented by the Indian Government. The issues established through this paper are ones that need to be recognized by academics, governments and implementation agencies, and considered in the implementation of post-disaster reconstruction programmes in the future.

1. Introduction

Post-disaster reconstruction (PDR) is a very complex process that requires multi-sectoral involvement, significant resourcing and a wide range of skills (da Silva, 2010). Different decisions need to be taken that can drastically alter the effectiveness and ultimate long-term vulnerability of a community. It is essential that PDR processes break the traditional disaster cycle by implementing mitigation measures during the recovery and reconstruction phase. Disasters themselves are devastating, but they do provide the opportunity for governments and communities to implement strategies and frameworks that not only recover and reconstruct but also mitigate against the consequences of further disasters (Davis, 1978).

It is essential that this reconstruction process adopts building techniques that are familiar to the affected benefactors. The use of familiar techniques will reduce the length of the project allowing people to improve their livelihoods as early as possible. The controlling government will ultimately make decisions that will dictate whether the reconstruction projects will produce short-term gains or reduce long-term vulnerability while improving sustainability and installing a legacy of disaster-risk reduction (DRR) (Thiruppugazh, 2010b).

It has been understood for a number of years by numerous academics that, for effective housing and shelter reconstruction after disasters, it is not only necessary to build more resilient housing, but the needs of the community must be considered (Davis, 1978; Twigg, 2000; Ruwanpura, 2008).

This paper will focus on the long-term consequences of different safe housing projects that were implemented after the Gujarat earthquake in 2001. The main objective of the research was to identify structural vulnerabilities that developed as a result of the different projects. Field research was conducted to identify trends and issues that have arisen as a result of different safe housing projects, which may adversely affect a community's vulnerability. Questionnaires were also conducted to understand the impact different PDR approaches have on reducing long-term vulnerability. Appropriate knowledge transfer has a direct relationship with developing a community's capacity to reduce their long-term vulnerability.

1.1. The 2001 Gujarat earthquake and its impact

On 26 January 2001 at 8:46 am an earthquake with a reported magnitude of 7.7 hit the Kachchh district in the northwest region of the state of Gujarat, India. The earthquake was felt throughout northwest India, Pakistan and even felt in southern Nepal and parts of Bangladesh (USGS, 2010). The earthquake reportedly killed 13,085 people with another 167,000 injured and requiring medical attention. The death toll and casualties sustained were primarily due to the substantial number of damaged and collapsed buildings totalling around 950,000 and 230,000, respectively. The majority of the effected buildings were located over 7,500 villages with a total of 450 villages being totally destroyed. Four major towns and cities were also badly damaged requiring extensive town planning during the reconstruction (Thiruppugazh, 2010a).

1.2. Government of Gujarat reconstruction policy

The government of Gujarat set up a governmental organization, the Gujarat State Disaster Mitigation Authority (GSDMA). Their goal was to oversee every aspect of the rehabilitation and reconstruction. The main aim of the GSDMA was to ensure a sustainable recovery while laying the foundations to improve livelihoods and build capacities for development and to introduce a vigorous disaster management scheme within Gujarat. The Gujarat Earthquake Emergency Relief Program was implemented after the earthquake and included four major principles.

- debris removal, salvage and recycling;
- construction of temporary shelter;
- reconstruction of more than 230,000 houses;
- repair and strengthening of over 1,000,000 houses.

The government's primary objective for re-housing was for a participatory, community-driven process with individual households and communities working on a self-help rebuilding basis. Technical support and guidance was to be provided by the government, the private sector and NGOs. The government's most important policy was to implement a community/owner-driven in-situ housing recovery plan. This approach also had backing from the World Bank as they viewed this method as the most appropriate for reducing vulnerability and assisting continued sustainability (World Bank, 2001).

1.3. Achievements, capacity building and DRR

In general, it was perceived that the reconstruction process after the 2001 Gujarat earthquake was very successful. It was reported that the PDR effort was the most expansive and fastest reconstruction and rehabilitation programme of its kind. By 2004, 98 per cent of the total 230,000 houses had been reconstructed. Seventy-five per cent of the completed houses were constructed through variations of the owner-driven reconstruction (ODR) approach. This ODR approach was the largest of its kind to be extensively backed by a government. The remaining 25 per cent were reconstructed through what is known as the donor-driven reconstruction (DDR) approach. It was deemed very successful that only 2 per cent of communities were fully relocated and with only 5 per cent being partially relocated (Thiruppugazh, 2010b). The reconstruction process received international recognition from both the UN and the World Bank for its initiatives and implementation. Through the reconstruction more than 29,000 masons and 6,000 engineers were trained in the design and construction of multi-hazard reconstruction (Mishra, 2008).

One aspect that the earthquake and ensuing damage brought to the fore was the deficit in the knowledge, practice and implementation of design standards in the state of Gujarat. Since the earthquake, a number of initiatives have been launched to improve seismic capacity building throughout the state. These initiatives include the development of seismic engineering design codes, upgrading practical skills and raising awareness levels in what has been described as a disorganized construction sector. The regulatory frameworks and monitoring mechanisms have also been improved, most notably concerning the improvement and implementation of the Indian design standards (Sheth et al., 2004).

2. Post-disaster reconstruction approaches

There are a number of different post-disaster reconstruction approaches adopted by governments after disasters. These processes are set out in a handbook by the World Bank 'Safer Homes Stronger Communities' which was published in 2010 nine years after the earthquake (SHSC, 2010). The different approaches primarily relate to the amount of control a household possesses over the reconstruction of their homes. The selection of a specific PDR approach should be decided on the basis of the results of a damage and loss assessment conducted after the disaster along with consultation with the affected communities and the lead disaster agency. In a PDR guide, five policies are presented; these are formulated as a result of PDR experience (EPC, 2004). It is suggested that these policies are key to the success of any reconstruction approach and they should include self-reliance, decentralization and empowerment, equality, mitigation and minimum relocation. The different approaches principally fall into two different reconstruction philosophies ODR and DDR.

2.1. Owner-driven reconstruction

The ODR approach allows the respondents to reconstruct their destroyed homes with financial assistance and technical guidance provided by the government. The respondents may undertake the construction work or repairs; but normally they will enlist the service of a trained family member or local contractor/artisans.

There are a number of advantages of the ODR approach that have been reported by numerous academics. These studies primarily comprise comparative studies of different reconstruction processes. The most tangible advantages relate to the overall cost being significantly lower. Other advantages relate to the owner having the option to move in prior to completion of the works, and with this the occupancy rates tend to be higher than in the DDR approach. There are also a number of intangible advantages, it is understood that reconstruction and specifically house design need to be sensitive to the respondents culture, religious views, social requirements and their understanding of the function of the house (Barakat, 2003). These considerations can be achieved through the ODR approach by giving the respondent an active participatory role during the construction of their home. This approach can also restore a sense of pride and well-being in the respondents and the community through their active participation. The main advantage of this approach allows respondents to build their homes according to their own preference and requirements, within certain parameters. This approach is also better placed to play a key role in building capacity within a community by assisting knowledge transfer in the event of additional constructions or extensions. It has also been said that this approach is best equipped, and most likely to preserve, the local architectural heritage and vernacular housing style (Barenstein, 2006).

There are also a number of issues that need to be considered when implementing this approach. Most importantly, this approach requires more commitment and involvement from the government. They need to train engineers and artisans to construct and oversee construction. They also need to be in control of material resourcing and financial funding to ensure materials are affordable. The building codes used during construction also need to be adaptable and not biased towards indifferent vernacular housing types that are common to a certain community/area. The approach could also present problems with the involvement of vulnerable groups, as they may not be in a position to manage the reconstruction (SHSC, 2010).

The GSDMA implemented compensation packages that were distributed based on the systematic surveys carried out by a team comprising of a government engineer, a local government representative and a representative from the local community. The level of compensation ranged between Rs. 40,000 and Rs. 90,000 depending on information collected during the surveys including housing type, size and level of damage. The financial assistance was distributed in three instalments, all at varying levels of construction. First, 40 per cent of the total compensation was paid during the preparatory stage. Second, another 40 per cent was paid on completion of the walls and the third payment of 20 per cent was paid on completion of the structure including a plaster finish. The second and third payments were only granted on verification and certification from government engineers. To aid this approach, the Gujarat government placed a civil engineer in every community to provide guidance and supervision during the construction phase (Barenstein, 2006).

2.2. Donor-driven reconstruction

The DDR approach is normally controlled by an external non-government agency on an individual community basis. This agency enlists the services of a contractor to undertake the design and construction works. The DDR approach is also commonly known as contractor or agency-driven reconstruction. This approach is primarily adopted by governments as it is considered to be the easiest and quickest way of providing housing on a large scale to affected communities and establishing normality after a disaster. The lead contractor is likely to source outside of the community when producing the design, acquiring materials and expertise (SHSC, 2010). This reconstruction approach can be conducted in situ or in a location that is deemed more appropriate.

There are a number of documented advantages of the DDR such as the speed and relative ease of implementation compared to other approaches. It has also been documented that the DDR can be preferred where there is a lack of construction knowledge and self-building traditions within a community. Lead agencies also consider this the most appropriate approach when it is deemed that the site conditions are too unsafe to reconstruct and the whole community needs to be relocated. Vulnerable beneficiaries also benefit as they may be incapacitated and unable to take advantage of the ODR approach. The DDR strategy can also be very effective and more appropriate when reconstructing dense urban populations, rental housing and buildings utilizing complex techniques (SHSC, 2010).

There are, however, a number of reported disadvantages with this approach. There are a number of reports that the DDR approach has an adverse effect on socioeconomic structure of the local communities and can drastically disrupt the recipient's livelihoods (Badri et al., 2006). It has also been reported that in some cases intended recipients have been reluctant to move into their new homes (Sanderson and Shama, 2006). Barenstein (2006) also reported problems relating to the poor quality of the constructed houses, primarily due to the inexperience of agencies and their understanding of construction techniques. Other issues, with this approach, relate to the loss of local building culture and capacity which can mean making extensions or repairs difficult and expensive to implement, as the knowledge and skills of the earthquake construction technique are not known by the occupant. There have also been difficulties when this approach was implemented in rural locations as the housing designs, layout and building technology are alien to the respondents (Karunasena and Rameezdeen, 2010).

3. Research methodology

The research methodologies were developed to answer two hypotheses.

First, that there are a number of structural vulnerability issues that arise as a result of different types of PDR approaches. These issues need to be identified and understood.

From the papers reviewed, a number discusses the relative merits of PDR approaches. There is no information relating to the long-term structural safety of the different reconstruction approaches. This ultimately affects a community's current vulnerability in the face of further earthquakes and, therefore, needs to be addressed.

Major issues and questions also need to be raised when considering how the quality of materials and construction would be regulated; especially when certain elements of the Indian design codes require highly trained artisans and engineers to ensure that the specific requirements are implemented correctly. It is also worth noting that through the framework set up by the GSDMA there is no stipulation for quality checks of materials at any stage in either the ODR or DDR approaches.

Second, the Indian government will struggle to incorporate effective technical knowledge transfer which would ultimately reduce a community's vulnerability and facilitate a legacy of DRR.

The PDR process in Gujarat was the first to extensively back the ODR approach on a large scale and it was deemed a massive success. One of the major advantages of the ODR approach is that it is capable of increasing the capacity of a community through effective knowledge transfer. There was a big shift in the common building practice in Gujarat after the earthquake, from adobe and unreinforced masonry (INCE, 2001) to confined masonry. If this transformation is to be replicated and a legacy of reduced vulnerability developed, not only is an entire construction workforce required to be retrained but the continued sourcing of raw materials also needs to be addressed. The shift in construction practice raises important questions relating to the additional financial cost of further construction and whether these costs can be afforded by the poor.

Prior to starting the field research serious concerns were raised regarding certain requirements specified in the Indian design standard. Most notably was their recommendation of the use of asbestos sheeting (Indian Standard, 2002).

Three different research methodologies were developed to investigate the different hypotheses.

3.1. Community leader/senior community member discussion

On entering a community the first task was to identify the community leader or a senior member of the community who were willing to be interviewed. The interviews were semi-structured. They initially started with a discussion to understand the dynamics and make-up of the community and how it was affected by the earthquake. Grand tour questions were then asked, which revolved first around the reconstruction effort within the community, with prompts to understand the involvement of the community members and their perception of the success or not of the reconstruction. A second grand tour question was then asked that concerned the community's current stance on DRR. Prompts to understand knowledge transfer, current building practice and whether the children of the community had been told about the earthquake were also approached.

3.2. Visual surveys

The surveys were completed and recorded using a visual survey recording sheet. The visual survey recording sheets were developed and created through reviewing a number of different earthquake-resistant building principles and assessing the Indian design standards. The structure and layout of the visual survey recording sheets were also produced from reviewing other surveying techniques and procedures, notably ATC-20, FEMA 310 and the Earthquake Engineering Field Investigation Team surveying guide (Hughes and Lubkowski, 2000).

The following construction principles were used to assess the earthquake resistance of the surveyed buildings, as presented by Coburn et al. (1995).

Principle one, Robust building form – This was recorded by visualizing the layout both inside and outside of the structure in the vertical and horizontal plane.

Principle two, Foundations – The foundation construction and depth were impossible to survey. However, the following were recorded to try and understand how a foundation would respond during an earthquake: the level of foundation in a horizontal plane ensuring there are no large variations, the width of the foundation to the width of the wall and finally by examining the overall quality and finish.

Principle three, Material quality – This was mainly assessed by identifying the quality of the materials used. An understanding of the mortar strength was gauged through touch. Where possible, concrete was inspected to identify any bubbling or inadequate finishing. Reinforcement was also checked to ensure that it was ribbed and no tension cracks had developed where it had been bent. Cracks within the plaster finish were also recorded especially where it was perceived that they were created from more than heat expansion and compression of materials.

Principle four, Walls – The strength and quality of the wall construction were assessed by checking that their alignment was square, the quality of the mortar and blockwork/brickwork were checked and whether substantial cracking was present within the plaster.

Principle five, Distributed openings – All structural openings including doors and windows were measured to ensure that they fulfilled specific criteria as set out in the Indian design standards (Indian Standard, 2002).

Principle six, Connections and ductility – This was recorded by identifying the connection details and material type between the foundations, walls and the roof, whether reinforced concrete, timber or confined and reinforced masonry.

Principle seven, Roof – The type of roofing system was recorded as well as the connection method used to connect the roofing systems to the walls.

Principle eight, Non-structural elements – All non-structural elements that could cause damage were recorded, with special attention being paid to any connection details.

3.3. Questionnaires

The questionnaire was developed in order to understand an individual's involvement during the construction of their home. Questions were orchestrated to understand the levels of technical knowledge transfer from engineers to beneficiaries especially considering the different reconstruction approaches. The questionnaire was organized into five different sections. Section 1 relates to an individual's involvement during the planning, design and construction phases. Section 2 relates to whether a family was required to relocate as part of the reconstruction process. Section 3 endeavoured to understand how adequate a beneficiary perceived the layout and design of their new house was when they moved in. Section 4 approached the issue of the level of both education and advice given by different construction agencies. The advice and education related to a beneficiaries knowledge and importance of the earthquake resistant features (ERF), the need for appropriate maintenance. It also related to the beneficiaries understanding of the specific requirements for altering the fabric of their homes along with the implications if this is incorrectly undertaken. Section 5 questions whether the recipients actively discuss the earthquake and its consequences with their children.

The sampling criterion was first defined by All India Disaster Mitigation Institute (AIDMI) who selected a variety of communities that were extensively affected by the earthquake. These communities were where AIDMI had played an active part in there development and rejuvenation since the earthquake. Within these communities, interviews were conducted as described above, following the interviews visual survey respondents were identified by conducting a reconnaissance of the community. This was to identify a cross representation of buildings reconstructed including adapted structures and buildings constructed by families of varying levels of affluence.

4. Results

Field research was conducted across six villages and one urban setting. The urban setting consisted of three different locations of varying affluence. The fieldwork was undertaken in two districts of Gujarat namely Kachchh and Patan for 10 days in July 2010, nine-and-a-half years after the devastating 2001 earthquake.

Five discussions with the community leader were conducted with 40 visual surveys of buildings and 43 respondents of the questionnaire.

The visual surveys highlighted a number of different issues that included:

Temporary shelters were only evident at locations where respondents had reconstructed in situ through the ODR approach. This accounted for 18 different sites. Out of these 18 sites, 50 per cent of the respondents were still occupying unsafe temporary shelter.

Extensions had been constructed at 12 out of the 40 structures surveyed. Out of these 12 surveys, six extensions were constructed by individuals who opted to reconstruct through the ODR approach with the remaining through the DDR approach. Out of the 12 extensions constructed only six incorporated ERF, with three from respondents who constructed via the ODR approach and three via the DDR approach. From the six extensions that incorporated ERF it was deemed that only two were satisfactorily constructed in accordance with the Indian design standards, ensuring life safety of its occupants.

Construction quality could only be gauged from 14 of the 40 buildings surveyed. This was due to the wide spread use of plaster as an architectural finish. Out of these 14 surveys, 12 were deemed to have elements of construction that were insufficient and would significantly impair the response of the building during an earthquake. Poor construction practices were also identified in a number of current construction projects.

A poor housing scheme set up by the government had housed families that had failed to incorporate ERF. Two different buildings were surveyed that had been constructed through this scheme.

Asbestos sheeting was identified in a quarter of the surveyed houses. This was where extensions and new constructions were built. It was noted that asbestos sheeting was not used as part of any of the safe housing projects reconstruction after the 2001 Gujarat earthquake. It was, however, evident that asbestos sheeting had been used by a recipient of the ODR construction scheme implemented by the government to re-house the homeless or poor. There is evidence to show that asbestos sheeting is used by both poor and middle-class individuals.

Non-structural elements in the form of high level shelving had been incorporated into the fabric of every surveyed building. This shelving was constructed with no guards or doors. Out of the 40 surveyed buildings, 16 were perceived as having the potential to cause serious injuries in the event of another earthquake as this shelving was used to hold heavy household goods.

Questionnaire results are shown in Table 1. In total, 43 questionnaires were completed, 40 were of respondents of the visual surveys and a further three questionnaires were completed of respondents who had reconstructed through the

TABLE 1 Comparison of ODR and DDR approaches response to the questionnaire

ODR approaches			DDR approaches		
Question	Yes	No	Question	Yes	No
1, Owner's input			1, Owner's input		
a, planning	23	4	a, planning	4	12
b, design	22	5	b, design	4	12
c, construction	24	3	c, construction	3	13
2, Relocation	8	19	2, Relocation	12	4
3, Adequately	21	6	3, Adequately	15	1
4, Structural awareness			4, Structural awareness		
a, features	22	5	a, features	5	11
b, importance	22	5	b, importance	5	11
c, maintenance	18	9	c, maintenance	3	13
d, altering	18	9	d, altering	2	14
e, implications	17	10	e, implications	2	14
5, Children	18	9	5, Children	2	14
Head of house	25	2	Head of house	15	1

owner-driven approach, where the weather restricted the completion of a visual survey.

By considering the different factors from both the visual survey and questionnaire answers vulnerability levels were assigned to the 40 completed surveys. Out of the total of 40 surveys completed, 17 of the respondents were deemed as living in or owning structures that had high vulnerability in the event of further earthquakes. Eleven surveyed respondents were deemed as being moderately vulnerable in the face of another earthquake and the remaining 12 were deemed as being at low risk in the event of further earthquakes.

5. Discussion

It is first worth discussing the statistical relevance of the results. In total, 40 visual surveys were completed across nine different communities. This is a small sample group when it is considered that around 230,000 homes in over 7,000 communities were reconstructed. However, what the results do show is that a number of common themes where evident across the different communities. The exact extent of these issues cannot be quantified but what is important is that these issues are highlighted so they can be considered by governments and implementation agencies when undertaking large-scale reconstruction programmes.

These key findings were derived from compiling the results for both the visual surveys and questionnaires. The major findings will now be discussed with specific relevance to how they relate to the hypotheses.

5.1. Temporary shelter

The continued habitation of temporary shelters constructed in the interim period during the completion of beneficiaries' homes was identified as a significant issue. This problem was originally highlighted by Jigyasu (2001) as early as 9 months after the earthquake. The temporary shelters surveyed were primarily constructed from irregular-shaped locally sourced stone bound by mortar consisting of an unknown sand cement mix refer to Figure 1. The roofing systems primarily consisted of a timber truss or joist system covered by either asbestos sheeting or clay tiles. The construction neglected the use of ERF including plinth, lintel and ring beams. Subsequently, this type of construction is very vulnerable in the event of further earthquakes. Only two temporary shelters were deemed as ensuring life safety and these had been constructed using mud and straw.

All the respondents who were still occupying their temporary shelters were deemed to be part of the middle or lower social economic group within the community. From the surveys it was also identified that the continued occupancy of unsafe temporary shelters related to the choice within a given community. This was particularly the case in two communities where it was evident that around 75 per cent of the affected families were still living in unsafe temporary shelters. It was also clear that respondents who

FIGURE 1 Occupation of unsafe temporary shelter.

had constructed these temporary shelters were aware that they lived in very unsafe temporary shelter. But, they were very reluctant to demolish what they had spent time, money and effort constructing. The main reason these houses were constructed unsafely is primarily due to respondents' lack of understanding of ERF at the time of construction, and financial constraints.

This is a major issue that relates to the in situ, ODR approach. Implementation agencies or government need to consider strategies that either provide the affected respondents with safe interim shelter or insist that as part of the ODR philosophy these unsafe temporary shelters are demolished.

5.2. Extensions

Understanding the construction of extensions provided an insight into the quality of knowledge transfer and would show if this knowledge transfer has been successful in establishing a legacy of long-term DRR. Figure 2 shows two extensions constructed on top of original DDR approaches. Both extensions were built with no connection to the existing structure, the original DDR building designs neglected to consider features that would allow benefactors to expand their original construction. There was no vertical steel protruding through the roof at column intersections that would allow a benefactor to construct an extension in accordance with design standards.

It is also worth considering some of the responses given to the questionnaire from the respondents who had constructed extensions. The results showed that there was a difference in the understanding about ERF between individuals who were involved during the construction of their homes and individuals who were not. Some of the reasons given for individuals who did not incorporate earthquake features included

FIGURE 2 Unsafe extensions neglecting ERF.

financial constraints, thinking that their inclusion was either unnecessary or that there were other more pressing problems that needed to be addressed before constructing to include ERF.

After the earthquake, a number of communities set up construction agencies and committees to oversee the construction of new buildings and extensions. These bodies were primarily set up to assist knowledge transfer and support long-term DRR within a community. Out of the 10 structures that were considered inadequate, nine were constructed in communities that had active construction committees. In one case the construction agencies only requested to see initial architectural drawing and then completed an inspection of the finished plastered structure. In another case where an extension was built, the respondents were only required to inform the external organization who was involved with the reconstruction.

It is evident from the results that individuals who were involved with the construction of their homes have a good understanding of the need to incorporate ERF, as also heighted by Thiruppugazh (2010a, b). But it is also evident that this knowledge transfer is not contributing to reducing the long-term vulnerability of the community. The results indicate that there are still issues surrounding effective knowledge transfer and safe construction practice along with their implementation.

The results also suggest that the reluctance or incorrect implementation of the Indian design standards could relate to the following: construction in India and other developing countries is primarily undertaken by the home owners themselves, whereas in developed countries these activities are completed largely by trained and competent individuals. Considering this, a significant issue could be related to home owners not possessing the understanding or skills to implement what is a very technical process normally completed by experienced and trained individuals. The financial cost to implement the Indian design standards could also be beyond the respondent especially if they do not possess the required skills to construct, and need to hire a trained mason.

If there is to be more of a emphasis on construction committees to oversee safe construction, it is evident that these bodies need further training and guidance on how and to what extent ERF needs to be incorporated. It is also evident that they need to incorporate a more efficient and effective checking/quality control process.

Another concerning issue that has developed as a result of inappropriate knowledge transfer is that the ODR approach has not been translated into an effective DRR mechanism.

5.3. Construction quality

The field research also highlighted a number of issues surrounding the quality of in-situ construction. The factors that were deemed inappropriate included extensive cracking of plaster, bubbling concrete, exposed reinforcement and insufficient amounts of mortar used to bond blockwork, refer to Figure 3.

With the extent and level of the poor construction practices identified, a concerning trend could have developed as a result of the ODR philosophy. This problem could also be a factor in the DDR approach. The final process of the ODR approach is to release the final payment of the completed structure after an inspection from a government engineer, and surprisingly this inspection relates to the plaster finish. First it must be asked to what extent is the plaster finish covering up insufficient and poor-quality construction especially considering the extent of poor practices identified. In one case a community leader believed that the quality of construction within his community was very poor. Another community leader also stated that they used the plastered architectural finish to mimic the presence of ERF that they had purposely neglected. The number one goal of any post-disaster reconstruction programme must be to reduce the vulnerability of the affected individuals; this can only be done by providing housing that is capable of withstanding hazards to which an

FIGURE 3 Inadequate quality of construction.

area is prone. The ODR process must focus on ensuring that the buildings are constructed adequately. By introducing a framework that comprises criteria that include architectural finishes the reconstruction process is focusing on elements that do not reduce benefactor's vulnerability in the event of further earthquakes.

Another alarming issue that was also identified relates to what was stated above. It became evident through various discussions that the inspection process neglected any kind of testing of the in-situ construction materials and that the release of payments through the ODR approach had no bearing on the quality of construction.

Poor construction quality is a very concerning issue. The results have identified that the Indian government have struggled to instil a legacy of effective knowledge transfer, especially when it is considered that the respondents have little or no understanding of quality assurance and good construction practices. Similar issues were raised during the reconstruction after the 2004 Indian Ocean earthquake by da Silva (2010) and Potangaroa (2010) who both discuss the problems surrounding quality control and the need for improved field testing.

5.4. Poor housing scheme

Another concern relates to a government scheme recently set up to house the poor and homeless. The scheme works under the same philosophy as the ODR approach where funds are released for the completion of certain elements of work. When asked why the owners had not included these features, the respondents stated that the money reserved by the government was not enough and it did not allow them to even consider including the ERF. It was also stated that a government engineer undertook checks just to ensure that construction works were being undertaken and not to check the inclusion of ERF or the quality of construction. The government of Gujarat needs to reconsider their scheme for housing the poor. It is underfunded and is subsequently allowing families to live in buildings that are very vulnerable in the event of another earthquake.

It is of much concern that the government have developed and introduced a scheme based on the ODR approach that shows total neglect for the beneficiary involved. For an effective technical knowledge transfer process to reduce long-term vulnerability, the government needs to include appropriate frameworks that provide earthquake resistant housing for the poor.

5.5. Asbestos sheeting

In India the use of asbestos sheeting is well accepted and even suggested in the Indian design standards as a possible construction material to be used as part of a roofing system (Indian Standard, 2002). This is of particular concern as asbestos has significant health implications including respiratory problems that lead to lung cancer. These health implications would be more significant when it is considered how asbestos sheeting would react during an earthquake. The dangerous fibres would break up causing their inhalation to become more wide spread. The common use of asbestos shows a significant lack of understanding on the part of the Indian government especially when considering its use has been outlawed in over 60 countries throughout the world.

What this shows is those individuals constructing using asbestos are complying with the Indian design standards and reducing their vulnerability to the consequences of structural failure. But, unknowingly making them open to other vulnerabilities that they are not aware of could pose a problem later in life. The Indian government needs to implement policies that prohibit the manufacturing and distribution of asbestos sheeting and provide a non-harmful asbestos alternative.

5.6. Non-structural elements

Non-structural elements can be classified in two categories: external and internal. From all 40 surveys there were no issues concerning the presence of external non-structural elements such as water tanks. There were, however, issues concerning the internal non-structural elements and specifically high-level shelving.

Governments and construction agencies need to have more of an understanding for non-structural elements. The inclusion of Cupboards or guards could significantly reduce the consequences of heavy non-structural elements falling from height, particularly in areas that are used as sleeping quarters.

5.7. Knowledge transfer and questionnaire

The results from the questionnaire indicate that there is direct link between a beneficiary who was involved during the construction of their homes and their knowledge of ERF.

The results also indicate that where people were involved and gained the knowledge of the ERF they also had an understanding of the importance of adequate maintenance, the need and methods required to ensure alterations and extensions are undertaken safely and what the implications could be if they alter the fabric of their home incorrectly. The result of the visual surveys have shown, however, that even though a number of respondents have this technical knowledge they still occupy buildings that are unsafe in the event of further earthquakes.

There is also an interesting trend that where individuals have extensive knowledge of the ERF they are more likely to discuss the earthquake and its consequences with their children.

It is also worth noting that there is no great discrepancy between the ODR and DDR approaches when considering the perception of the respondents to the adequacy of the design and layout of their new homes when they moved in.

6. Conclusion

This research has highlighted a number of issues that have arisen as a result of the reconstruction after the 2001 Gujarat Earthquake. The findings should be considered in the event of implementing further PDR projects. The findings fall into two distinct categories. First, the research has highlighted a number of themes and issues that need to be considered when implementing any post-disaster reconstruction programme. Second, there are a number of findings that indicate that the Indian government needs to consider a number of their current policies to facilitate a lasting legacy of DRR.

There were restrictions to the research; the sample size was relatively small with only 40 surveys completed across nine communities. Constraints on the findings and results also related to actually being able to comprehend the vulnerability of the surveyed structures. This was due to the extensive use of plaster. Further physical testing would also be required to fully understand the quality of in-situ construction.

The ODR approach is considered by many as the most effective reconstruction approach, but there are still a number of concerns that need to be considered during its implementation. One factor is the continued habitation of unsafe temporary shelter constructed in the near vicinity of their new homes. Another factor is the significance given to the architectural plastered finish to release the final payment as part of the ODR framework, especially considering the level of poor construction practices that were identified.

The idea of creating a PDR approach that will instil a legacy of DRR through technical knowledge transfer could be complicated. Constructing to comply with design codes and standards throughout the world can be a difficult exercise, especially considering the financial constraints and access to raw materials. Its implementation requires guidance from trained engineers with the construction being undertaken by experienced artisan's proficient in the construction of earthquake-resistant housing. Gujarat is an interesting case, the post-disaster reconstructing approach set out to introduce a construction practice that was alien to the majority of its population. This has led to significant issues regarding quality and the continued misunderstanding of earthquake-resistant construction. To instil a legacy of DRR, a reconstruction programme not only needs to train a construction industry in the new practise but also ensure raw materials are available at an affordable price.

Another major issue concerns the construction of extensions. In nearly every case where extensions were constructed by recipients of both the ODR and DDR approaches the constructions would be unsafe in the event of further earthquakes, even when in some cases earthquake-resistant features had been incorporated. There also needs to be greater consideration so that benefactors can safely construct extensions on their original constructed homes, specifically concerning the DDR approach.

The results from the questionnaire showed that there is certainly a higher level of knowledge transfer when individuals are involved during the construction of their own homes. But this knowledge transfer has not created a legacy of DRR.

The government of Gujarat also needs to consider at least two of their current policies. First, relates to the scheme set up to re-house the poor and homeless. The scheme is significantly underfunded and causing respondents to construct and live in houses that are very vulnerable to earthquakes. Second, relates to their approval and recommendation through the Indian design standards of the construction material asbestos.

Out of the total of 40 surveys completed, 17 of the respondents who were surveyed were deemed as living in or owning structures that had high vulnerability in the event of further earthquakes. Eleven surveyed respondents were deemed as being moderately vulnerable in the face of another earthquake and the remaining 12 were deemed as being at low risk in the event of further earthquakes.

In the light of evidence presented in this paper from the fieldwork conducted, a number of issues were highlighted that should be considered when implementing PDR projects in the future. Not only does the current vulnerability of respondents need to be managed and reduced but communities also need in-depth assistance in long-term DRR.

Acknowledgements

This research paper would not have been possible without the assistants from the All India Disaster Mitigation Institution. Their knowledge, experience and reputation within the state of Gujarat ensured that the field research produced viable outcomes. I would also like to thank my supervisors at University College London for their support, especially John Twigg for his technical assistance and links within the humanitarian sector.

References

ATC-20. *Detailed Evaluation Safety Assessment Form [online].* Applied Technology Council. www.atcouncil.org/pdfs/DETAIL.PDF (accessed 19 August 2010).

Badri, S. A., Asgary, A., Eftekari, A. R. and Levy, J., 2006. Post-disaster resettlement, development and change: a case study of the 1990 Manjil earthquake in Iran. *Oversees Development Institute, Disasters*, 30(4). 451–468.

Barakat, S., 2003. *Housing Reconstruction after Conflict and Disaster*. Humanitarian Practise Network Paper, Number 43.

Barenstein, J. D., 2006. *Housing Reconstruction in Post-earthquake Gujarat [online]*. Humanitarian Practice Network, Report Number 54. www.odihpn.org/report.asp?id=2782.

Coburn, A., Spence, R. J. S., Pomonis, A. and Hughes, R., 1995. *Technical Principles of Building Safety*. Intermediate Technology Publication, London.

da Silva, J., 2010. *Lessons from Aceh, Key Considerations in Post-disaster Reconstruction*. Practical Action Publishing Ltd., Rugby, Warwickshire.

Davis, I., 1978. *Shelter after Disaster*. Oxford Polytechnic Press, Oxford.

EPC, 2004. *Participatory Planning Guide for Post-disaster Reconstruction [online]*. Environmental Planning Collaborative. www.tcgillc.com/tcgidocs/TCGI%20Disaster%20Guide.pdf (accessed 21 August 2010).

FEMA, 310. *Handbook for the Seismic Evaluation of Buildings [online]*. US Department for Home Security. www.fema.gov/library/viewRecord.do?id=1534 (accessed 19 August 2010).

Hughes, R. and Lubkowski, Z. A., 2000. *The Survey of Earthquake Damaged Non Engineering Structure [online]*. Earthquake Engineers Field Investigation Team. www.istructe.org/eefit/files/field.pdf (accessed 20 December 2010).

INCE, 2001. *Indian Census, Table of Houses [online]*. Government of India. www.censusindia.gov.in/Census_Data_2001/Census_data_finder/Census_Data_Finder.aspx (accessed 15 July 2010).

Indian Standard, 2002. IS 4326: 1993 Edition 3.2 2002, *Earthquake-resistant Design and Construction of Buildings – Code of Practise*. New Delhi.

Jigyasu, 2001. *Post-Earthquake Rehabilitation in Gujarat – 9 Months After [online]*. Radical Interpretations of Disaster. www.radixonline.org/gujarat4.htm (accessed 21 August 2010).

Karunasena, G. and Rameezdeen, R., 2010. Post-disaster housing reconstruction: Comparative study of donor vs ODR approaches. *International Journal of Disaster Resilience in the Built Environment*, 1(2). 173–191.

Mishra, P. K., 2008. *The Gujarat Earthquake Recovery and Reconstruction Experiance 2001–2004 [online]*. Power Point Presentation. www.un.org.cn/public/resource/d848821f04ab0313c0ac1ac590bc39dd.pdf (accessed 10 July 2010).

Potangaroa, R., 2010. *5th I-Rec Conference and Workshop. Structural Control in the Field*, Presentation, 16 July 2010, CEPT University, Ahmadabad, India.

Ruwanpura, K. N., 2008. Putting houses in place: rebuilding communities in post-tsunami Sri Lanka. *Cities*, 21(6). 527–536.

Sanderson, D. and Shama, A., 2006. Winners and losers from the 2001 Gujarat earthquake. *Environment and Urbanisation*, 20(1). 177–186.

Sheth, A., Sudhir, K. J. and Thiruppugazh, V., 2004. Earthquake capacity building and risk reduction measures in Gujarat post Bhuj 2001 earthquake. *13th World Conference on Earthquake Engineering*, 1–6 August 2004. Vancouver, Canada. Paper No, 2018.

SHSC, 2010. *Stronger Homes Safer Communities, A Handbook for Reconstruction after Natural Disasters [online]*. The World Bank. www.housingreconstruction.org/housing/ (accessed 21 August 2010).

Thiruppugazh, 2010a. *I-Rec Conference and Workshop. Gujarat Earthquake Reconstruction Problems, Performance and Possibilities*, 16 August 2010. CEPT University, Ahmadabad, India.

Thiruppugazh, 2010b. *I-Rec Conference and Workshop. NGO Participation in Reconstruction: Knowledge Transfer and Capacity Building for Sustainability: A Case Study of Gujarat*, 17 August 2010. CEPT University, Ahmadabad, India.

Twigg, J., 2000. The age of accountability? Further community involvement in disaster reduction. *The Australian Journal of Emergency Management*, 18(2). 51–58.

USGS, 2010. *Preliminary Earthquake Report [online]*. United States Geological Services. http://neic.usgs.gov/neis/eq_depot/2001/eq_010126/ (accessed 30 July 2010).

World Bank, 2001. *Gujarat Earthquake Recovery Program, Assessment Report [online]*. Prevention Web. www.preventionweb.net/files/2608_fullreport.pdf (accessed 19 August 2010).

Resilient dwellings or resilient people? Towards people-centred reconstruction

THEO SCHILDERMAN[1] AND MICHAL LYONS[2]

[1]Practical Action, International Policy and Programmes, The Schumacher Centre for Technology and Development, Bourton on Dunsmore, Rugby, CV23 9QZ, UK

[2]Social and Policy Studies, Faculty of Arts and Human Sciences, London South Bank University, London, UK

Our review of post-disaster reconstruction in 10 countries has demonstrated some of the weaknesses in currently dominant approaches, e.g. donor-driven reconstruction and owner-driven reconstruction. They have often been top-down and exclusionary, focusing on people with existing title to land and housing, and failing to reach the marginalized, especially in urban areas. Since they aim for safer dwellings, they rarely tackle people s underlying vulnerabilities. This paper argues that an inclusive and participatory approach, people-centred reconstruction, should be central to housing and livelihoods after disasters. Reconstruction programmes need to make people more resilient to future risks. That requires not just making their buildings safer, but also making people more capable to adapt to risk. As to housing, many agencies interpret building back is better, as reconstructed houses are safer than pre-disaster types. That concern for quality leads them to set high standards, engage architects and engineers to produce designs, and use contractors to construct. The end product is often inappropriate, difficult to maintain and too expensive to replicate. Damage assessments after disasters often point at vernacular technologies such as timber frames that have performed much better than others. Provided if any weaknesses are addressed, they can be incorporated in reconstruction strategeies because they are well known to local residents and builders and use mainly local resources, they require less support, thus they are cheaper and quicker. The reduction of people's vulnerabilities, however, requires more than better housing; programmes also need to rebuild people's livelihoods, restore local markets and social networks. To strengthen their capabilities to cope, survivors should play key roles in decision-making and resource management. Forty years ago, John Turner concluded that the process of housing matters as much as its end product, as it empowers people. Reconstruction is not different: putting people at its centre empowers them and strengthens their capabilities and resilience.

1. Introduction

Over the past decade, the world has suffered a number of disasters of an unprecedented scale, including the Indian Ocean tsunami (26 December 2004), and earthquakes in Kashmir (8 October 2005), China (12 May 2008) and Haxti (12 January 2010). These required responses that exceeded those known after previous disasters. They also brought new actors onto the reconstruction scene, encouraged innovation and needed the scaling up of what previously had only been piloted. The authors were involved in the design, support and evaluation of some of the ensuing reconstruction projects, and learned valuable lessons in that process. From 2007, they collaborated with the International Federation of Red Cross and Red Crescent Societies (IFRC) on a more systematic assessment of large-scale reconstruction over the past two decades, which took nearly three years to implement and document. This included a survey of the literature and key project and programme evaluations, as well as in-depth guided studies by local researchers of large reconstruction programmes and projects in 10 countries.

We found that the reconstruction sector is changing only slowly. Many of the agencies involved are reluctant to move from a supply-driven relief mode to a supportive mode that is more appropriate to reconstruction. Often also, reconstruction takes place in isolation from the housing context and ignores livelihood issues. But there has been some innovation too, particularly since the turn of the century, often having a positive impact. This allows us to set out a way forward for reconstruction that puts affected people more at the centre and enables to better reach the poor.

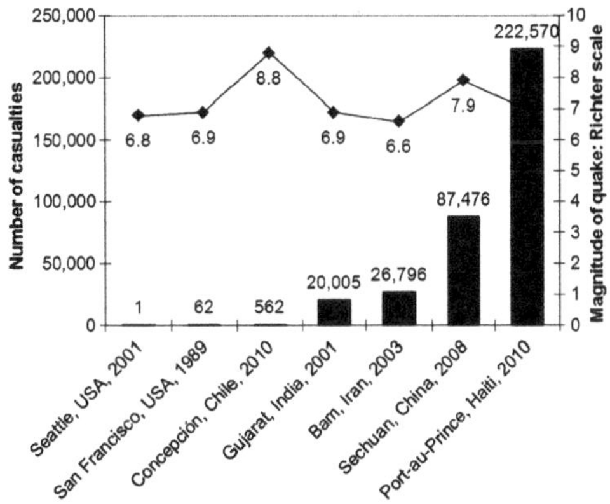

FIGURE 1 Earthquake magnitude versus casualties in different economies (EM-DAT database, November 2010).

2. The poor are more vulnerable to disasters

Over the past four decades, natural disasters have become more frequent and have affected growing numbers of people (EM-DAT database at Centre for the Research of the Epidemiology of Disasters). This trend is set to continue into the future, as a result of for example climate change, urbanization and persistent poverty. The increase in hydrometereological disasters, which is the category most influenced by climate change, is particularly significant. EM-DAT graphs also suggest a rise in people affected by disasters from around 45 million per year in around 1970 to 230 million around 2005.

It has been recognized for nearly 30 years that a clear link exists between development, vulnerability and disasters (Wijkman and Timberlake, 1984). As Figure 1 shows, disasters of similar magnitude have a much greater impact on poor countries than on rich ones. Within countries too, the poor are generally more affected than the rich. Over the past two decades, earthquakes of similar magnitude hit urban areas in the USA, India, Iran and Haxti, causing very few casualties in the USA, over 20,000 in India and Iran, and about 10 times as many in Haxti, by far the poorest country of these four. The slums and low-income settlements of Port-au-Prince were particularly badly affected. But when one of the strongest earthquakes measured ever (8.8 on the Richter scale) hit Concepción in Chile, just a few months after the earthquake of Port-au-Prince, the number of victims was relatively small.

Rich countries like the USA and Chile can afford to build to high-quality standards that incorporate disaster resistance. Other countries, such as India, have adequate construction standards, but inequalities within their population mean that many people cannot afford them. In poor countries like Haxti, over half the urban population cannot afford to build according to prevailing standards (Yahya et al., 2001). They construct houses as best they can, but their quality is often inadequate to sufficiently withstand disasters. Apart from lacking the money or knowledge and information to build better, the poor also often have to make do with dangerous sites to build on, for example steep slopes at risk of landslides, or alluvial plains at risk of flooding or liquefaction.

The majority of low-income housing in developing countries is built by the poor themselves and vulnerable to disasters. If nothing is done to overcome the above constraints, those vulnerabilities will persist in reconstruction. Besides, the livelihoods of the poor are also often more vulnerable to disasters and the poor have fewer resources to recover their livelihoods. Thus, poverty compounds the effects of disaster.

3. Reconstruction does not always make the poor less vulnerable

In the aftermath of disasters, there is an influx of resources, initially for relief, but then increasingly for recovery and reconstruction. This may not work equally well for everyone, and the poor are more likely to lose out. Support for housing reconstruction is currently often targeted at people who can prove their title to land or buildings, who are often better off. It discriminates against tenants and squatters, who frequently constitute the majority of the urban poor in developing countries, and against those who may have lost their titles, or have only traditional property rights. For instance, it is estimated that at least three-quarters of the urban population affected by the 2007 Pisco earthquake in Peru did have insufficient proof of property to benefit from a government reconstruction grant directed at house owners (Ferradas et al., 2011). This was a major reason for this cash-for-shelter programme to be abandoned.

Sometimes also, disasters can be used as an excuse by those in power to relocate poor people to so-called safer locations, far from livelihoods opportunities and their social networks. After the 2004 tsunami in Sri Lanka, housing for people living in a buffer zone along the coast was often built many kilometres inland; this had profound effects on the livelihoods of, for example, the fishermen among them. If they continued to work as fishermen, the distance from the sea placed an enormous burden of cost and time on this livelihood. The alternative, taking up different livelihoods and entering new markets required significant and sustained external investment in training, equipment and support over time. Housing reconstruction mostly takes place before reconstruction of livelihoods and often ignores its needs. If reconstruction programmes do not do enough to rebuild livelihoods, the poor may well end up worse than before.

Vulnerability can also be exacerbated because of local rivalries and unequal access to reconstruction funds. When money starts flowing in, it is frequently captured by elites, people who have the influence, connections and knowledge to deal with the bureaucracies in place or who may act as gate keepers and rent seekers. Even if, immediately after the disaster, the poor and the rich were working together towards recovery, this may change as soon as serious resources come in, and the poor may lose their position and influence. Thus, at the end of the reconstruction process, many poor people can be more vulnerable than before the disaster struck. Experience in Sri Lanka suggests that this trend can be mitigated through the institution of an accessible and proactive monitoring body which is independent of reconstruction programme management and funding (Lyons, 2009).

4. There is a disconnect between reconstruction and housing

Many agencies involved in reconstruction see it as a one-off activity that stands on its own; they fail to look at its link with housing. In essence, though, reconstruction deals with no more than a sudden upsurge in demand for housing, caused by a disaster. That may require some help, but not necessarily a different approach. Yet, after the tsunami hit Sri Lanka, for instance, the government set up a central body to deal with the reconstruction of around 120,000 houses (TAFREN, later replaced by RADA), rather than relying on the National Housing Authority which had a track record of supporting the construction of over a million houses in preceding decades with community participation.

It is important to develop a good understanding of the housing sector in a disaster location, when preparing for reconstruction, for four reasons:

1. Housing takes place in a context defined by policies, strategies, finance mechanisms, rules and regulations, culture, climate, knowledge and skills, available land and materials. Much of that context also applies to reconstruction, although a reconstruction strategy

sometimes changes parts of it, such as access to finance, or standards.

2. Understanding how and why people built as they did before a disaster can highlight underlying vulnerabilities and capabilities, as well as strengths and weaknesses in construction. Reconstruction can build on the strengths, but has to address the weaknesses. Some of the most successful reconstruction projects and programmes adopted vernacular skills and technologies with a good disaster record, sometimes incorporating moderate improvements. During the Kashmir earthquake of 2005, for instance, it was observed that traditional timber frame (*Dhajji Dewari*) walls performed well. Although this technology was in decline before the earthquake, with perhaps not more than 5,000 dwellings remaining, it was revived during reconstruction, and more than 100,000 of such houses were built anew (Stephenson, 2008) (Figure 2).
3. There is an important link between housing and livelihoods. This applies, for instance to its construction itself, with many people making a living from building, producing materials, providing transport, etc. Houses also often have livelihood functions, for example as a store, shop or workshop. And if people do not make their living from home, they do want to live in a location that is close to livelihood opportunities. Unfortunately, many reconstruction projects forget these links, and therefore do not do enough to rebuild livelihoods and the local economy.
4. The residents themselves play key roles in realizing housing. People who supported housing in the informal sector, such as Turner (1976), realized as early as the 1970s how important that participation is in strengthening their capabilities and empowering them. Following the Habitat Conference of 1976, this became a cornerstone of housing policies and strategies advocated by UN-Habitat and others. Some NGOs realized that the role of local actors is equally important in disaster mitigation and reconstruction, and that the processes involved can make them more resilient (see e.g. Maskrey, 1989).

FIGURE 2 Woman building a timber frame house after a recent earthquake in Chincha, Peru.

An analysis of housing generally should distinguish formal from informal housing, and rural from urban locations, in order to acknowledge the important differences in how housing is produced accordingly. Formal housing tends to be mostly urban, designed by professionals and built with durable materials according to prevailing rules and regulations, mostly by contractors. Because of this, it is more expensive, and therefore the typical housing of medium- to high-income groups. It is less vulnerable to disasters than the other types, but this is not guaranteed as disaster risks are often poorly understood, and standards can be circumvented. Informal housing occurs both in the slums and informal settlements of towns and cities, and in the countryside. It is often built incrementally by the residents themselves or local artisans, using materials that sometimes have a limited lifespan, or are recycled. Such housing generally does not use building professionals, nor apply prevailing rules and regulations. But there are important differences between informal housing in rural and urban contexts. Rural housing tends to use more naturally available materials, such as timber, thatch, stone or earth; it often has a rich tradition, with designs and technologies that have been passed on and improved over generations. Some of these may derive from previous experiences with disasters and are not to be ignored, as they can be quite resistant to disasters and considered for reconstruction. This happened with timber-framed housing in for example Pakistan, Turkey and Peru.

Informal urban housing tends to be the most disaster-prone of all. There are a number of factors contributing to this, for example that it is difficult for the poor to access safe sites. Sometimes they even prefer to live on relatively unsafe sites, because they are closer to livelihoods opportunities, and immediate survival can be a greater concern to them than an eventual disaster risk. In many developing countries, building standards are unaffordable to the poor (Yahya et al., 2001). And the procedures involved in building a house legally can be very cumbersome and costly, as described by de Soto for Peru (1989). There is also a real risk in some towns and cities for informal housing to be demolished, which is no incentive for investing in it. Finally, a lot of the urban poor are actually tenants, and landlords are often more concerned with a quick return on their investment than with tenant safety. This makes reconstruction for and with the poor in urban areas particularly challenging. It also makes learning from slum upgrading initiatives and housing projects with the urban poor implemented by organizations such as Slum/Shack Dwellers International even more relevant and urgent (Figure 3).

FIGURE 3 A two-storey timber frame house such as this one near the source of the Amazon in Peru may resist earthquakes and provide safety during flooding.

5. Approaches to reconstruction are changing only slowly

The first large reconstruction programmes to receive major external support took place after major earthquakes affected both Peru and Turkey in 1970. The approaches followed by governments and donor agencies at the time were to build houses for people rather than with them, often using alien designs and technologies. These were based on the principle that proper engineering and higher standards could solve the lack of resistance to disasters. They frequently involved relocation to sites that were considered safer but away from people's livelihoods. Several projects had important flaws, and as a result many houses remained unoccupied, and people affected reverted to their previous ways of building, remaining vulnerable to future risks (see e.g. Aysan and Oliver, 1987; Blaikie et al., 1994). They often also struggled to maintain or expand the houses, because the technologies were alien to them. This approach to reconstruction is often termed as *donor-driven reconstruction (DDR)*. It has been much studied since 1970, and the general conclusion of those studies is that DDR has too many drawbacks to be recommended, except in situations where very little local building capacity remains. These drawbacks include: (a) a lack of user participation, leading to inappropriate designs; (b) a preference for building many uniform houses on large sites, which are difficult and time consuming to acquire and service; (c) relocation away from livelihood opportunities; (d) high cost, but often with limited spin-off into the local economy; (e) lack of local capacity building; and (f) occasional poor quality. Unfortunately, notwithstanding the lessons learned, DDR still occurs to date, and even now some houses remain unoccupied (Figure 4).

Many humanitarian agencies that are traditionally involved in relief and have moved into reconstruction find it difficult to change from a supply-driven 'relief mode' to a much more supportive and participatory 'reconstruction mode'. Disasters also do put a lot of pressure on key decision makers. When many people are living in tents or makeshift accommodation, and the media put on additional pressure, decisions are frequently made in a hurry, without much investigation or analysis, and without involving all those that matter. After the 2004 tsunami, the Sri Lankan government, for instance, quite quickly established a buffer zone, within which the construction of houses was not allowed. The width of this zone ranged from 100 m above the high-level water line in the south and west of the island, to 200 m in the east and 500 m in areas in the north under LTTE control. But this meant that half the affected population would have to be relocated, and it was difficult to find enough land to do so. Under pressure from civil society and aid agencies, the government was forced to reduce the zone a year after the tsunami happened, to 35–125 m (Hidellage and Usoof, 2010) (Figure 5).

Owner-driven reconstruction (ODR) has existed even longer than DDR, because it is the fall-back mode when people do not receive external assistance. NGOs have provided support to it for many decades, especially in Latin America, but usually on a small scale. Since about a decade, however, it has been attracting wider support, also from larger agencies, which has allowed it to scale up, especially in South Asia. This move perhaps mirrors the change in housing policies and practice, from supply driven to support driven, initiated at the 1976 Habitat Conference and implemented from there on. An increasing number of agencies is now querying why, when home owners play a major role in the production of houses under normal circumstances, reconstruction after disasters should happen in such a different way. In ODR, there is a bigger role for plot owners, with agencies playing a more supportive role. Most of the time, reconstruction happens on the original plot, allowing owners to make use of existing infrastructure which speeds up the process. Comparative evaluations of ODR and DDR, for example, in Sri Lanka and India, have shown that ODR is generally more successful

FIGURE 4 A DRR housing scheme built after the tsunami inland from Hambantota in Sri Lanka, away from livelihoods.

FIGURE 5 Houses built in Matara, supported by NGOs, in a post-tsunami ODR programme in Sri Lanka.

than DDR: has greater owner satisfaction; is quicker; is cheaper for agencies as owners add their own resources; strengthens social capital; and is better at incorporating livelihoods (see e.g. Duyne Barenstein, 2006; Lyons, 2007).

ODR is now becoming more widespread, and that has given rise to a number of variations in how agencies conceive the approach, in particularly when it comes to who is in the driving seat: agencies or owners. In some cases, agencies merely leave the building to the owners, but take the major decisions with regard to e.g. design, technology and financing themselves, perhaps with some consultation. Owner driven is not synonymous with owner built, although most ODR projects will adopt owner-built housing. But one recorded project of post-tsunami reconstruction in Aceh, for instance, changed from being owner-built to contractor-built housing, because the owners lacked essential building skills. They remained, however, involved in all key decisions on the project, which therefore remained largely owner driven (Da Silva and Batchelor, 2010). ODR has generally been more successful where agencies were prepared to leave more of the decision making to the owners. But it also has weaknesses, the main one of which is that it focuses on legal owners only, and is therefore an exclusive approach that does not cater adequately for the most marginalized. This has been particularly problematic in reconstruction in towns and cities, where there are relatively more non-owners. There also have been problems with its quality being inadequate, especially where agencies have provided too little technical support and capacity building. And standards are sometimes set too high, which carries the risk that the poorest may not be able to maintain that level in the long run, and become vulnerable again.

Another approach that has come to the fore in the last decade is that of *transitional housing*. It does not deliver permanent housing solutions, as DDR and ODR do, but is meant to be an interim solution between emergency shelter and the reconstruction of permanent housing. This helps to overcome the problem of those affected having to stay too long in emergency shelter such as tents, and at the same time provides some breathing space to better prepare permanent housing, with the involvement of those affected. Some agencies have a mandate that is limited to relief, which do not allow them to construct permanent housing, considered to be a development activity; however, they can fund transitional shelter, and actively do so. Those critical of transitional housing argue that it takes funding away from permanent reconstruction and recovery. But perhaps the biggest problem is not that, but the fact that transitional and permanent housing are not planned as a continuum, often because it involves different agencies. If transitional housing could be planned so that it either can grow into permanent housing, or that its components could be re-used for permanent housing, the problem of wasting resources and money on it would be reduced. As informal housing is very often incremental, such an approach might also be quite acceptable to those affected. However, as yet, there are few examples of it; one of those occurred in Kenya, and is further explained under Section 7 (Aubrey, 2010).

6. Towards people-centred reconstruction (PCR)

It is very tempting for humanitarian agencies to directly provide disaster victims with what agencies consider safer housing. After all, nobody would like to see similar disasters occurring in the same location in the future. Most houses built under DDR are certainly safer than the ones that collapsed. In ODR too, owners make an effort to build good quality; however, whether they are able to achieve that depends to an extent on the amount of technical support and training provided by supporting agencies.

But does either approach really reduce the vulnerability of all those affected by disasters in the longer term? What happens if they want to expand their house, or if their children get married and need to build a house of their own; will these new initiatives be equally safe? For that to happen, it is crucial that reconstruction approaches do reach everybody and are replicable by them. DDR frequently fails in that respect on several fronts: it sets high standards that are unaffordable, introduces alien designs and technologies, and does little to build capacity of residents and local builders. By not addressing livelihoods, it also does nothing to make its solutions more affordable in the future. There is ample evidence, from studies many years after DDR projects took place, of people reverting to pre-disaster housing designs and technologies, thus maintaining their old vulnerabilities.

ODR, in turn, does not reach many of the people affected, who will therefore remain vulnerable. Its solutions tend to be somewhat more affordable than DDR, and those affected will gain some skills and knowledge in the process, but how much depends entirely on agencies. As housing is mostly rebuilt in the same location, livelihoods opportunities may remain; however, projects frequently do far too little to help people restart livelihoods; in some cases, this has led to people diverting reconstruction funds to livelihoods activities, with a negative impact on reconstruction quality. As ODR has only happened at scale over the past decade, there has not been that much ad-post assessment of their housing situation and livelihoods many years after a disaster. So the judgment is still out as to how far ODR has helped to considerably reduce vulnerabilities. ODR is quite a broad term that is interpreted widely by agencies; the answer to this question is therefore unlikely to be uniform.

In their reconstruction programmes, agencies should look further than the delivery of safe houses. If they want to avoid hazards becoming disasters again in the future, such programmes should aim to make people themselves more resilient to risk and change. That requires considering not just the rebuilding of houses, but also of livelihoods, local markets and social capital. Above all, it needs the affected people to get involved in reconstruction processes, as these processes help to empower them and make them more resilient. We have called this approach *PCR*. PCR builds on some of the positive aspects of ODR but, by putting people at the centre, it also makes two important differences: (i) it focuses on all people affected by disasters, not just property owners, and (ii) it recognizes that a true reduction in vulnerabilities comes from making people, and not just their buildings, more resilient to shocks and change.

In order to achieve PCR at scale, our recent studies indicate that the following seven components are particularly important:

1. learning from the housing sector;
2. assessing what makes people vulnerable;
3. people participating from an early stage;
4. including all people;
5. empowering people;
6. incorporating livelihoods;
7. building back better.

7. Learning from the housing sector

PCR recognizes that residents play a major role in producing housing in normal circumstances and intends to apply what is good about that to reconstruction. In order to do so, stakeholders involved in reconstruction do need to gain a thorough understanding of pre-disaster housing. It is particularly important to find out how land is accessed and registered, how housing is financed, what rules and regulations apply and whether these are followed, what people value in the design, what the preferred technologies are and whether they incorporate any disaster-resistant features, who gets involved in building and what their skill levels are.

While it is tempting to aim for the reconstruction of complete houses of an adequate size and standard, the reality of normal housing is that it is often realized in an incremental process, with both the size and quality of the house improving over time. If there are inadequate resources to reconstruct complete houses for all those affected, a reconstruction strategy could build on this reality, by offering, for example, core housing that residents can expand later on, or incompletely finished housing. This approach has been taken up in, for example, Sri Lanka, Turkey and El Salvador (Lyons and Schilderman (eds, 2010) Chapters 4, 11 and 12). It can also be applied to make a deliberate attempt to create a continuum between transitional shelter and permanent housing, to save resources. An interesting example of that occurred in reconstruction with people affected by election violence in Kenya in December 2007 (Aubrey, 2010). Although this left half-a-million people displaced, there was little government funding or external aid available. Government was only able to offer approximately €100 as compensation to affected households, plus about €250 to those whose house was completely destroyed. Yet, it initially insisted on rebuilding three-bedroom brick houses. However, with the agricultural season starting, it was urgent to relocate those who could closer to their fields. And it was agreed that this could be done through transitional shelter. Another problem was that there was occasionally some uncertainty as to whether the locations people were going to resettle in would remain secure. The shelter cluster then formed a working group of NGOs that developed a joint shelter strategy and concept design, that was subsequently piloted by the NGO GOAL, in partnership with the Office of the United Nations High Commissioner for Refugees. The working group based the strategy and design on the following principles:

- rapid construction not to lose too much time in the high season for agriculture;
- potential to upgrade transitional shelters to permanent housing;
- ability to disassemble the shelter and rebuild it elsewhere;
- ability to disassemble it and re-use its components for permanent housing;
- potential to expand from basic SPHERE standards in line with owner needs.

The result was a shelter of 3 × 6 m to provide space for five people. It used timber poles as frame, covered by corrugated galvanized iron sheets. Beneficiaries could use any locally available materials for the walls. The floor was made of soil, but raised above a ground level that was sloping away from the house (Figure 6).

FIGURE 6 A transitional shelter in Kenya has been upgraded using timber off-cuts.

8. Assessing what makes people vulnerable

When a moderate earthquake struck the Alto Mayo region of Peru in 1990, the region was in decline. With agriculture being the main source of income, farmers were facing problems marketing their main product, rice, because the government has disbanded the agency buying rice from farmers, and failed to maintain the one major road that linked the region to the markets of the main cities on the coast. Incomes had declined, which affected people's ability to build and maintain their houses well. This proved to be a major factor in the damage and casualties the earthquake caused. The Alto Mayo's inhabitants had become more vulnerable as a result of their livelihoods declining due to external factors. To make things worse, when aid started to flow into the region, it included a lot of imported rice, at a time when local stores were full to the brim with rice that farmers were unable to sell. This further worsened their potential for recovery, as it now became nearly impossible to sell rice locally. Damage observations after the earthquakes, however, showed that not everyone was equally affected. Immigrants into the region from the Andes had often built houses with heavy rammed earth (*tapial*) walls, the main technology they were familiar with, but it performed badly. A much lighter local technology, using timber poles, canes and mud infill (*quincha*), performed a lot better, with damage mainly occurring where walls had deteriorated by insect attack or humidity. This would subsequently form the basis for a reconstruction project by the NGO Practical Action, then known as Intermediate Technology Development Group (Schilderman, 1994).

In order to be able to build back better, those involved in reconstruction need to understand why the hazard that occurred became so disastrous to the people if affected. As the Alto Mayo case shows, that requires not just looking at the vulnerabilities of prevailing construction technologies, but also at the underlying reasons for those, which lie in people's own vulnerabilities, in this case declining incomes and insufficient knowledge of more appropriate technologies. Vulnerabilities and capabilities (or assets) are key components of people's livelihoods, and we need to understand people's livelihood strategies in a broader sense, to sometimes understand where and why vulnerabilities arise. A 1980s study of the Karakoram region of Northern Pakistan quoted by Twigg (no date), for instance, found houses to be dangerously located on steep slopes. Their owners were aware of the risks this posed, but preferred to build there rather than using the little flat and arable land for housing. Similarly, the urban poor often settle on more centrally located but dangerous sites, because these are closer to income-generating opportunities, rather than going for safer land on the fringes of towns. Immediate survival often is of greater concern to the poor than a potentially distant hazard; unless we understand these constraints, it becomes difficult to reconstruct sustainably.

9. People participating at an early stage

Comparative evaluations of reconstruction projects have found that people who had considerable participation in reconstruction ended up being more satisfied than those who were given houses (Duyne Barenstein, 2006; Lyons, 2007). The main reason for that is that, by being more involved in decision making, they felt they had more ownership. That does not surprise people with knowledge of housing, such as Turner, who concluded much the same in the 1970s. He argued in particular that it is the process of housing that is actually more important for people than the end product, since it builds their capacities and empowers them. But housing agencies often struggled to put participation into practice. Turner put the dilemma they were facing into one basic question: 'whose participation in whose decisions and whose actions?'. By now, there is ample evidence that participation and the establishment of effective partnerships between various stakeholders is a good approach to solve deficiencies in housing and related services, while at the same time building the social and human capital of those involved (Hamdi, 1995, 2010).

Yet, to date, many reconstruction agencies are still struggling with Turner's dilemma. Their definition of participation varies widely, with some restricting it to a mere participation in construction, as a means to cut costs, rather than an attempt to hand over some power to those affected. At the other end of the scale, there are agencies who allow households to decide on a wide range of matters, and have been able to do so at scale. The Community Recovery and Reconstruction Partnership (CRRP) in Sri Lanka was an initiative of the IFRC and UN-Habitat, with involvement of the SLRC, reaching some 11,000 households. It linked reconstruction to long-term development and vulnerability reduction, and included elements of capacity building and livelihood support. In terms of housing, the programme adhered to core government standards, but remained responsive to household preferences. It thus allowed for great diversity of plan, design and finishing, which contributed greatly to user satisfaction (Lyons, 2010) (Figure 7).

Assessing participation in a series of reconstruction projects in Peru, Guzmán Negrón

FIGURE 7 One of a large variety of houses built under the CRRP in Batticaloa, Sri Lanka.

(2010) concludes that even in a country where community participation is well embedded in tradition (*ayni*), it can at times be undermined by institutions that value product over process. This has happened in a number of cases where the church, the government or local politicians adopted a top-down approach, handing out food, equipment, services or housing support, without community participation. As a result, in some places people have become more dependent, and traditional forms of participation like the *ayni* are disappearing. In the long run, that may make them more vulnerable.

10. Including all people

There is evidence of several types of exclusion in recent reconstruction programmes, and it often involves people who were already among the most vulnerable before the disaster, for example, the low-income or landless; they risk ending up even more marginalized after reconstruction. In ODR, those who cannot prove they owned a plot or house before the disaster, quite often are not entitled to support. However, the mere fact that much financial assistance arrives can sometimes encourage or enforce change. Governments in Pakistan and Sri Lanka, for instance, have supported access to land by tenants and squatters through the release of public land or land purchase grants. And in Aceh, the lack of secure tenure, or proof of it, was overcome by a community land adjudication process (Lyons and Schilderman, 2010, Chapters 4,5 and 6).

The weak may also be excluded from the benefits of reconstruction if local elites capture power in those projects. Most of the projects studied had built in some safe-guards against that happening, but they were often only partial. In many cases, representative structures at community level did not require particular constituencies to be represented, and thus were often dominated by local leaders. Community meetings in Muslim areas were often not attended by women, who therefore had only indirect influence on projects. In other cases, though, construction was organized on a group basis, which allowed assigning marginalized households to groups with stronger households, and ensured that they did not fall behind as much as might have been expected.

11. Empowering people

Taking charge of a housing process can empower those involved (Turner, 1976); the same applies to reconstruction. People who are empowered become more resilient. Agencies, however, have to realize that just handing control to people, without offering the knowledge they need to properly manage a housing process, is hardly empowering. Duyne Barenstein and Iyengar (2010) describe how India has a long history of providing people with cash to build their houses, either in social housing or reconstruction programmes. Those programmes often have specifications that people are expected to follow, which do not conform to their traditional knowledge and skills. But they are not given technical support or training to make up for that, and sometimes the funding offered cannot meet the requirements, all of which can be really disempowering. Among the recent large reconstruction programmes in India, only the one after the 2001 earthquake in Gujarat combined cash grants with other measures to empower people, for example, a choice of options and adequate technical support. But a similar approach was neither adopted in Tamil Nadu after the 2004 tsunami, nor in Kashmir after the 2005 earthquake, because it was not institutionalized in policy. In a PCR approach, it is important for agencies to provide adequate technical support and training.

Social mobilization is as important as technical support in PCR. Social networks are an important community asset, and key in reducing vulnerability. Where they are strong, people will have traditions of mutual aid and information sharing. They are very important in the relief phase, but also in reconstruction. But disasters can weaken them, if key members are killed or get dispersed. Reconstruction programmes should therefore aim to re-establish and strengthen such networks.

Finally, people should have somewhere to go, if they have grievances, but this rarely happens. An exception was Sri Lanka after the tsunami, where the government established the Disaster Relief Monitoring Unit under the Sri Lanka Human Rights Commission as an appeals mechanism for individuals who were concerned about their treatment by state actors, and was able to identify and correct procedural errors and instances of corruption. This was supported by a campaign of rights education, and claims were handled by an independent ombudsman. RADA also established district grievance committees and a couple of major projects, such as the CRRP and Cash for Repair and Reconstruction (project), established their own grievance mechanisms, but being internal, they were considered less efficient (Hidellage and Usoof, 2010).

12. Incorporating livelihoods

The root cause of much vulnerability and suffering caused by disasters is poverty. Even in normal circumstances, it has been argued by for example Development Alternatives (2005, p. 100) that for housing to improve and become more disaster resistant, it would be important to strengthen the livelihoods of residents for them to be able to afford improvements. This can be enhanced by designing integrated projects aiming to improve housing and livelihoods, and restore local markets and social capital. How housing is designed and built in itself has livelihood implications, as it can be done in a way to enhance the local construction sector, and to incorporate residents' livelihoods needs in the design. Relocation of affected people should only be considered in extreme circumstances, as it tends to reduce their livelihood opportunities. Several DDR projects after the tsunami in Sri Lanka, for instance, near Hambantota, offered relocation inland, quite far away from the sea, which was totally inappropriate to the affected firsherfolk.

Unfortunately, most reconstruction programmes so far involve little holistic thinking. Most projects studied prioritized housing over livelihoods support, and most of the few that incorporated the latter did so at a very late stage. The unfortunate consequence of that is that some people will use cash meant for housing to restart their livelihood activities and then struggle to complete their house according to the required standards. A positive exception is a reconstruction programme managed by the Coffee Growers Federation after a 1999 earthquake in Colombia, described by Lizarralde (2010). This was a well-established federation used to provide various types of support to coffee growers before the earthquake, but with no experience in construction. This set-up functioned well to provide relief, and when the time came for reconstruction, the government reconstruction agency, Fondo de Reconstruccion del Eje Cafetero, invited the federation as a partner in rural reconstruction (Figure 8). They decided that households would be responsible for reconstruction, but hired in a team of engineers to support that process, and took care to provide adequate information. They set up a fund which offered a basic grant of $4,000 for rebuilding a house, with the option of an additional loan of $1,000. The fund also offered a grant to rebuild coffee infrastructure of $2,000, with the option

FIGURE 8 Rebuilt house with coffee-processing plant at the rear in Colombia.

of a further $3,000 loan. This combination of housing and livelihood support proved to be very effective in practice.

13. Building back better

Many agencies aim to 'build back better', and define it as reconstructing houses that are more resistant than pre-disaster types. That concern for quality often leads them to hire architects and engineers to make designs, and seek contractors to construct, which in most cases is equivalent to DDR. We have argued above, though, that people can take charge of their own reconstruction processes and achieve decent quality, provided they receive adequate support and training to address known weaknesses.

It is also important for agencies to consider what the long-term impact may be of the ways of reconstruction they are promoting. Standards for reconstruction tend to vary considerably between disasters and across projects, often as a direct consequence of available aid. If standards are set too high after a disaster, because aid is available, there is a real risk they may not be maintained in the future. In Sri Lanka, for instance, we found that the standards set by government for reconstruction, were significantly higher, both in size and quality, than the housing most affected people were used to before the tsunami. They required a minimum size of 45 m^2 and the use of durable materials; a catalogue of reconstruction projects produced by the Tsunami Housing Support Project (GTZ, 2006), shows a wide range in costs, with the average working out at 885,000 Rps, or about $9,000 per unit. If standards are set too high, there is a real risk they will not be replicated by a sizeable population when new building has to occur in the future, and that will leave them vulnerable.

A proper assessment of local housing and the damage caused by a disaster (Sections 7 and 8), however, can direct stakeholders to vernacular designs and technologies that have proven in practice to be relatively more resistant. If they have minor weaknesses, they can often be overcome with little extra effort and money. Using locally known building methods does reduce the amount of technical support and training needed, and will boost the local economy. And there is evidence, for example, in the case of *improved quincha* in Peru or *dhajji dewari* in Pakistan that those improved vernacular technologies are the ones most likely to be taken up at scale.

14. Conclusion

Past approaches to reconstruction have often been too narrowly focused on providing disaster victims with safer houses. Evidence from 40 years of reconstruction aid has shown that this has often been expensive, inappropriate and unsustainable. We argue in this paper that a new approach to reconstruction is needed which looks beyond the end product of a resilient house and aims to increase the overall resilience of the affected people. It needs to be much more integrated and place those people at the centre. Such an approach can learn much from the housing sector, where it has been recognized for a long time that being involved in the process is as important to empower people and make them more resilient, as the end product. And this approach needs to consider not just the rebuilding of houses, but also of livelihoods, local markets and social networks.

References

Aubrey, D., 2010. *Kenya: Can Temporary Shelter Contribute to Participatory Reconstruction?* Practical Action Publishing, Rugby, UK.

Aysan, Y. and Oliver, P., 1987. *Housing and Culture after Earthquakes: A Guide for Future Policy Making on Housing in Seismic Areas*. Oxford Polytechnic, Oxford, UK.

Blaikie, P., Cannon, T., Davis, I. and Wisner, B., 1994. *At Risk: Natural Hazards, People's Vulnerability and Disasters*. Routledge, London, UK.

Da Silva, J. and Batchelor, V., 2010. *Indonesia: Understanding Agency Policy in a National Context*. Practical Action Publishing, Rugby, UK.

De Soto, H., 1989. *The Other Path: The Invisible Revolution in the Third World*. I.B. Tauris & Co. Ltd., London.

Development Alternatives, 2005. *Participatory Rural Habitat Processes: Emerging Trends*. Development Alternatives, New Delhi, India.

Duyne Barenstein, J., 2006. *Housing Reconstruction in Post Earthquake Gujarat: A Comparative Analysis*. Humanitarian Policy Group Network Paper 54, ODI, London, UK.

Duyne Barenstein, J. and Iyengar, S., 2010. *India, From a Culture of Housing to a Philosophy of Reconstruction*. Practical Action Publishing, Rugby, UK.

EM-DAT, the OFDA/CRED, 2010. *Emergency Events Data Base*. www.emdat.be/natural-disasters-trends.

Ferradas, P., Schilderman, T., Vilela, A. and Mariscal, J., 2011. *Participative Reconstruction and Risk Reduction in Ica*. WRC Innovation Competition Entry. The World Bank, Washington, DC.

GTZ, 2006. *Towards Sustainability: Building Practices in Post-Tsunami Housing Programmes*. Tsunami Housing Support Project, Colombo.

Guzmán Negrón, E., 2010. *Peru, the Long-Term Impact of Short-Term Reconstruction Work*. Practical Action Publishing, Rugby, UK.

Hamdi, N., 1995. *Housing without Houses: Participation, Flexibility, Enablement*. Intermediate Technology Publications, London, UK.

Hamdi, N., 2010. *The Placemaker's Guide to Building Community*. Earthscan Publications, London, UK.

Hidellage, V. and Usoof, A., 2010. *Scaling-up People-Centred Reconstruction: Lessons from Sri Lanka's Post-Tsunami Owner-Driven Programme*. Practical Action Publishing, Rugby, UK.

Lizarralde, G., 2010. *Decentralizing (re)Construction: Agriculture Co-Operatives as a Vehicle for Reconstruction in Colombia*. Practical Action Publishing, Rugby, UK.

Lyons, M., 2007. *Building Back Better: Large-scale Impacts of Small-scale Approaches to Reconstruction*. IDEAR Working Paper 2007/1, London South Bank University, London, UK.

Lyons, M., 2009. Building back better: on the large-scale impact of small-scale approaches to post-disaster reconstruction. *World Development*, 37(2). 385–398.

Lyons, M., 2010. *Can large-Scale Participation be People-Centred? Evaluating Reconstruction As Development*. Practical Action Publishing, Rugby, UK.

Lyons, M. and Schilderman, T. (eds), 2010. *Building Back Better, Delivering People-centred Housing Reconstruction at Scale*. Practical Action Publishing, Rugby, UK. www.practicalaction.org/reconstruction.

Maskrey, A., 1989. *El Manejo Popular de los Desastres Naturales: Estudios de vulnerabilidad y mitigactión*. ITDG, Lima, Peru.

Schilderman, T., 1994. Adapting traditional shelter for disaster mitigation and reconstruction: experiences with community-based approaches. *Building Research & Information*, 35(5). 414–426.

Stephenson, M., 2008. Notes from experience in post-earthquake rural housing reconstruction in Pakistan. Presented at the Building Back Better Workshop, Bejing, China. www.un.org.cn/public/resource/9330387be56a506bac9cae9aef6d5400.pdf.

Turner, J., 1976. *Housing by People: Towards Autonomy in Building Environments*. Marion Boyars Publishers Ltd., London, UK.

Twigg, J., no date. *'Disasters, Development and Vulnerability', Development at Risk?* UK National Committee for the IDNDR, London. 2–7.

Wijkman, A. and Timberlake, L., 1984. *Natural Disasters: Acts of God or Acts of Man?* Earthscan, London, UK.

Yahya, S., Agevi, E., Lowe, L., Mugova, A., Musandu-Nyamayaro, O. and Schilderman, T., 2001. *Double Standards, Single Purpose: Reforming Housing Regulations to Reduce Poverty*. ITDG Publishing, London, UK.

Community-led resettlement: From a flood-affected slum to a new society in Pune, India

VICTORIA CRONIN AND PETER GUTHRIE

Department of Engineering, Centre for Sustainable Development, University of Cambridge, Trumpington Street, Cambridge CB2 1PZ, UK

This paper describes the resettlement process of a community devastated by annual floods, to newly constructed housing in Pune, India. The relocation from Kamgar Putala slum to a housing society at Hadapsar was organized by a community-led NGO partnership in 2004. The housing development was coordinated by the local NGO Shelter Associates with significant community participation. The housing has been revisited in 2010 to evaluate the sustainability of the resettlement project's delivery model via stakeholder perception. The process of organizing for resettlement after natural disaster is described along with the implementation and evaluation of the new housing nearly six years after initial occupation. The strong partnership approach overcame a series of political and financial hurdles at various stages of the relocation project. The story of resettling Kamgar Putala is detailed alongside an outline of the current political climate for an alternative slum-upgrading policy in India and Pune. The advantages of an empowered community supported by an influential local NGO demonstrate a commendable team effort which has tackled the threat of floods. The paper highlights the merits of a community-led partnership approach to housing development for achieving sustainable urban development as well as the alleviation of poverty in a developing context.

1. Introduction

An estimated 1 billion people live in urban slums and this number is increasing by 25 million each year; this means that more than half of the world's poor will live in cities by 2035 (United Nations Department of Economic and Social Affairs, 2006). A slum or informal settlement is characterized as substandard housing, lacking access to clean water, sanitation and other infrastructure, overcrowding, insecure tenure, unhealthy and hazardous conditions, poverty and social exclusion (United Nations Human Settlements Programme, 2003). In the past poverty was thought to be a rural issue, but the locus of world poverty is now shifting from the rural to the urban context (OCHA/IRIN and UN-Habitat, 2007). New slums are forming and existing slums are growing. In poor countries slums make up 30–70 per cent of urban populations (United Nations Population Fund, 2007). Poor institutional and financial resilience, less robust infrastructure, rapid industrialization, a reliance on natural resources, strain of population growth and urbanization are some of the causes that lead to the formation of slums.

The urban poor are often the most vulnerable to natural disasters as they inhabit overcrowded, marginal, unstable and dangerous land with no financial cushion or security. When flooding, fire, earthquakes, landslides and cyclones strike, the urban poor are often the worse hit. The circumstances of the urban poor also make them more vulnerable to dangers such as organized crime and epidemics (OCHA/IRIN and UN-Habitat, 2007). In India, many slums are often located in urban locations that are particularly vulnerable and upon sites deemed not developable for habitable buildings; such as hill-slopes, railway tracks, pavements and by water bodies. As is the case for the settlement of Kamgar Putala in Pune, it is common for slums to be situated on river banks which are susceptible to rising water levels particularly in the monsoon season. As many slums are made of impermanent materials and poorly constructed, these floods can have a disastrous effect on residents and their homes. Water damage, complete destruction and the related health impacts from the spread of waterborne diseases all have devastating consequences.

Approaches to tackle slums can involve physical, social, in-situ and resettlement options. There are numerous housing delivery systems with adaptive and proactive measures. Current good practice experience shows that community participation is vital to the long-term success of slum-upgrading projects. One example of effective community participation in the delivery of a slum-upgrading project brought about by flooding is the relocation of Kamgar Putala to Hadapsar in Pune, India.

2. Initiating resettlement

The informal settlement of Kamgar Putala is located along the banks of the river Mutha at Shivaji Nagar in Pune and is vulnerable due to the threat of road widening and flooding. Every year the monsoon rains cause this stretch of the river to flood into the settlement which is constructed from temporary materials. In August 1997, the rains were particularly devastating resulting in one-third of the entire settlement being submerged (Shelter Associates, 2001a, b) as well as significant damage to other slum pockets situated on the river edge. Coupled with municipality plans to widen the adjacent Sangam Bridge it was not sustainable for the residents of Kamgar Putala to remain living at Shivaji Nagar.

After the severe 1997 flood-water had subsided, a comprehensive survey of the six main affected areas along the river was conducted in 1998, as requested by the Pune Municipal Corporation (PMC). The PMC specified a survey of all the settlements situated along the river Mutha between Kamgar Putala and Mhatre bridge with the aim to assess how many houses were affected by the floods. This survey was undertaken by collectives of Pune slum dwellers Mahila Milan, together with the National Slum Dwellers Federation of Mumbai, and the NGOs Shelter Associates and SPARC. The community federations used this opportunity to initiate the process of bringing communities together to negotiate with the city authorities for the development of a resettlement strategy (SPARC & Shelter Associates, 1998).

Shelter Associates is an NGO comprising architects and social workers who work with community federations. The main focus of their work is housing and infrastructure for the poor. The guiding philosophy of their work is that 'poor people have the best solutions to their problems and are quite capable of not only devising strategies which work best for them, but also implementing them' (SPARC & Shelter Associates, 1998). It was in this context that the NGOs supported the community federations to initiate their own relocation strategy to move away from the dangers of living by the river and to be involved in the development of their own new housing society at a new site in Hadapsar.

3. The flood

The largest flood in the history of Kamgar Putala came on 23 August 1997. Six slum pockets are situated along the banks of the river Mutha between Kamgar Putala and Dattawadi. The water came between 5.30 a.m. and 10.00 a.m. submerging 847, nearly half of all the houses. The residents had been accustomed to flooding every monsoon affecting the houses on the lower banks; however in 1997, the floods were disastrous. The water came while people were sleeping and residents had not received any prior warning. Some immediate relief was provided by local authorities, schools, residents and youth groups who tried to arrange for temporary food and shelter. The affected people camped in schools, the stock exchange building, community halls and the garages of nearby housing complexes. The water started to recede after 4 days, but many people were not able to gain access to their houses until 15 days after the flood (SPARC & Shelter Associates, 1998).

In the settlement of Kamgar Putala, 156 houses were completely swept away by the river. Four hundred and sixty-eight houses (31 per cent of all) were submerged. In other houses the water had left silt, rubbish, snakes, insects and sewerage that caused terrible damage. The insects, reptiles and damp caused many children to become ill. For immediate relief, the authorities arranged for some money, a few utensils and clothing to be given to the flood-affected people. The distribution of the relief was based on ration cards. Unfortunately, many deserving people missed out as their documents were lost in the flood. The people whose homes were completely destroyed were forced to rebuild their houses *pucca*, in brick or *patra* despite the increased cost (SPARC & Shelter Associates, 1998).

At the time of the survey many residents had been living at Kamgar Putala for 16 years. They favoured the location due to its convenience for local schools, hospitals, transportation, markets and work places, despite the danger of the flooding river nearby. The favourable city centre location might also have been preferred by residents due to a lack of other available locations within which to settle. Despite the hazardous location many residents may have considered the proximity to livelihoods more important than the threat of flooding (Figure 1).

FIGURE 1 The devastating floods at Kamgar Putala ©Shelter Associates.

4. Road widening

As well as the dangers of the water, the residents of Kamgar Putala were also at threat of eviction from road widening. The municipality had plans to widen the Sangam Bridge adjacent to the settlement. When the 1997 flood line and the boundaries of the proposed road widening were marked on a map, only 25 per cent of the original settlement was not affected. The NGOs realized that the communal facilities of the entire settlement would be affected as well as valuable community networks and relationships among families, friends and neighbours if the settlement were to be split. Therefore, any resettlement option needed to include everyone (Shelter Associates, 2001a, b) (Figures 2 and 3).

5. The survey and community mobilization

The PMC started by appointing surveyors to mark the high flood line across the six main settlements affected by the floods. The line indicated all the houses that had been either partially or fully submerged. Next, a team of 20 people from the Mumbai Mahila Milan and Mumbai NSDF came to Pune to conduct a comprehensive survey of the flood-affected people. These people joined forces with a further 40 members of the Pune Mahila Milan to conduct the survey together. The survey process involved numbering all the houses then, a door-to-door enumeration, community meetings and the creation of settlement registers. The survey created a great deal of energy among the residents of Kamgar Putala. Some were curious and excited, whereas others were wary and scared of the threat of eviction. The survey process heightened the awareness of Mahila Milan enabling the federation to gain

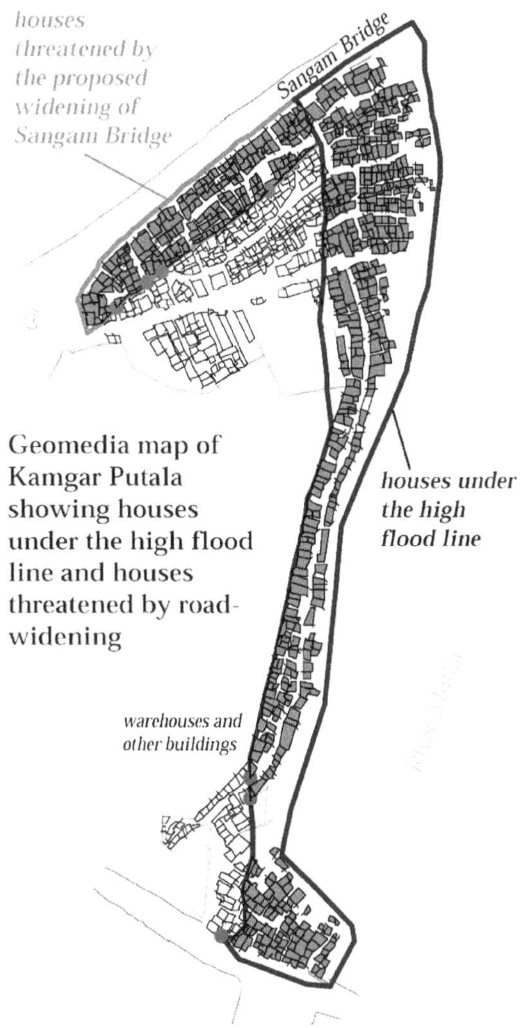

FIGURE 2 Geomedia map of Kamgar Putala (Shelter Associates, 2001a, b).

more members (SPARC & Shelter Associates, 1998).

The survey conducted in Kamgar Putala in January 1998 found it to be a declared settlement (officially recognized by the authorities) home to approximately 3,768 people. The slum was said to be established in 1947 and in 1998 constituted 928 houses of various construction ranging from *pucca* houses on the upper banks, *kutcha* and semi-*pucca* on the lower banks. Large amounts of demographic data of all the six

FIGURE 3 After slum clearing at Kamgar Putala.

affected settlements were collected during the survey. The majority of the respondents were Maharashtrians, but a large proportion of the population comprised second- or third-generation migrants. In all, 57.1 per cent were born in Pune city. The largest group of non-Maharashtrians came from Uttar Pradesh. A total of 71.5 per cent of residents were said to have lived in the settlement for over 10 years, 49.5 per cent for over 20 years and 10.5 per cent for more than 5 years. The remaining 10 per cent could not recall when they came to the settlement (Shelter Associates, 2001a, b) (SPARC & Shelter Associates, 1998). These data show that 82 per cent of residents have been living at their present address for over 5 years and prior to 1995 which qualifies them for resettlement according to government regulations. Most of the houses were sized between 90 and 140 ft^2 the small size indicates the level of congested living conditions. The survey also revealed the number of joint and extended families sharing one house.

The surveyors found that many families had lost their ration cards and photo ID documentation during the floods. These documents are required to prove residency status and therefore eligibility for resettlement under government schemes. The tracing and reissuing of such documents is a long task and implies that data gathered during the survey will need to be continuously updated and maintained (SPARC & Shelter Associates, 1998).

Throughout the survey process, Shelter Associates and Mahila Milan held meetings with the residents and motivated them to join the federation activities. Residents were encouraged to begin crisis savings groups for future use when the inevitable resettlement came. At the time of community consultation there was an ongoing resettlement project at Dattawadi further up the river, the success and public awareness of this project helped Mahila Milan gain the confidence of Kamgar Putala's resident's that resettlement could also be a viable option for them. Resistance did come from some people who were scared to move to a new location and also some traditional and political leaders. The research conducted by Shelter Associates found that many residents were so accustomed to battling against the floods each year that it had become a way of life, from which many people were resistant to change. However, the positive impact of the Dattawadi relocation generated much interest and people began to realize that there was a viable alternative to their situation. The enthused residents began housing saving schemes and became actively involved in land searches while participating in the Mahila Milan federation activities. Negotiations to acquire a relocation site began and the mobilization of the community had begun. The newly formed federations started to develop strategies for resettlement and to take control of their own future (SPARC & Shelter Associates, 1998).

6. Baandhani and shelter associates

After the surveys were complete, the women of Kamgar Putala got together to start daily saving schemes and to hold regular community meetings forming the community-based organization (CBO) Baandhani. This women's collective was also instrumental in the 1998 River Workshop that was held at the Dattawadi relocation project site with an aim to raise the profile of the slum dwellers that were being overseen by the PMC. The PMC had initiated a river improvement project with a team of advising experts, but the riverside slums and key slum resident stakeholders were not included in the consultations. At the River Workshop women from the slums spoke about the hazards of their life by the river and expressed their desire to be involved in any potential resettlement schemes (Shelter Associates, 2001a, b).

The women of Baandhani started up housing savings groups to ensure that they had the funds for a loan down payment if and when the time came that resettlement became a reality. While awaiting resettlement, the collective were also very active in endeavours to improve the conditions of their present environment by initiating vermiculture activities (funded by the PMC) to decompose their biodegradable waste and produce manure to sell thereby generating an income. This work attracted the attention of the Additional Municipal Commissioner Deepak Kapoor (1999) who visited their project and expressed his interest in the work of the collective (Shelter Associates, 2004).

The women negotiated with the PMC to demonstrate the difficulties they were facing with living at Kamgar Putala and their willingness and capability to contribute to their own resettlement scheme. The community successfully managed to raise their profile, thanks to the support of the NGO Shelter Associates that worked to strengthen the voice of the poor and mobilize the community collectives (Shelter Associates, 2001a, b).

7. Governance and planning

Due to the institutional structure of local government in Pune the voice of the Kamgar Putala community was sometimes missed and the municipality had other priorities. During the period of 1997–2002 there were many transfers and changes within the PMC and administration heads. Each time a new civic chief started the Baandhani federation was forced to restart their negotiations to convince the administration to prioritize their resettlement plans.

In 2000, the efforts of the Baandhani federation were noticed by the Municipal Commissioner Ratnaker Gaikwad. He visited the community and accepted invitations to meetings with Shelter Associates who had tentatively prepared housing plans for the resettlement to present to him. These meetings were successful and the commissioner agreed to support the project. With the PMC's support a 2.4 ha plot of flat land reserved for the economically weaker section (EWS) at Hadapsar was identified as a location for the residents of Kamgar Putala to resettle to; a safe plot far away from the water. There was, however, some disagreement among local politicians who campaigned for the community to be re-housed in the same place, in situ, with plans to construct a 5 crore rupees retaining wall to keep the water out. Their justification for this was that the majority of residents were rag-pickers and scrap vendors who relied on this income which was easily accessible by train from the Kamgar Putala site. The politicians did not feel that the residents would want to move away from their livelihoods (Shelter Associates, 2004). However, Baandhani and Shelter Associates disagreed with this, so they set about proving the facts. After Shelter Associates' household surveys, which included employment patterns of the residents, it was clear to see that the majority of workers were in fact unskilled construction labourers who did not actually require a specific site for their place of work. The other employment groups identified, skilled workers, auto drivers, shop owners informed Shelter Associates, that moving was not harmful to their trade. The residents claimed that even if it were an issue they would be willing to make this sacrifice to move to more secure homes away from the threat of floods (Shelter Associates, 2004). Shelter Associates research also found that the majority of female workers were domestic maids and the organization was confident that the Hadapsar site was centrally positioned enough to provide employment locally.

Due to uncoordinated government development planning for the city of Pune, Shelter Associate's plans and designs for the resettlement scheme needed much iteration and numerous proposal submissions to the PMC (Sen et al., 2003) (Table 1).

The uncoordinated planning can be further demonstrated by the fact that while the relocation plans were being discussed with the PMC, another PMC department was constructing two large community toilet blocks and a community

TABLE 1 Iteration of development plan

Plan 1 1997–99: After the 1997 floods the PMC drew a high flood line across the settlement. After surveys it was clear that 379 houses would be affected by the floods. At this point SA and Baandhani started designing a development plan for the 379 households on the Hadapsar site
Plan 2: In 1999 a newly formed Mutha River Improvement Committee comprising prominent city activists, environmentalists and scholars proposed the construction of a 'submersible road' in the riverbed which required more houses to be cleared. This meant that a new total of 559 households would need to be relocated and SA had to revise their plans
Plan 3: After the second revision of plans were submitted to the PMC, SA then learned that the Sangam Railway Bridge running over Kamgar Putala was to be widened by 20 m. This increased the total number of affected households to 793. As the total number of houses in the settlement was 1014 the PMC suggested that the entire settlement should now be relocated which meant the project had tripled in size. After ihe third set of designs had been submitted to the PMC the Building Department then realised that part of the existing settlement had been reserved for in situ redevelopment. With the support of GIS mapping (Sen et al., 2003) SA demonstrated that half of the previously reserved land was taken up by bridge widening, road building and the high flood area. Shelter Associates submitted a feasibility report to the PMC to show that it would not be possible to re-house the entire settlement in situ on what was left of the previously reserved land. It would only be possible to re-house 240 families in an 11 storey building, the rest of the community would be stranded. SA had proved that resettlement was the only viable solution for Kamgar Pula's residents

centre within the areas of the slum threatened with demolition (Shelter Associates, 2004).

Despite the planning submissions and successful negotiation so far between Baandhani, Shelter Associates and the PMC everything changed in 2001 when the head of the government administration, at that time, changed again. The resettlement plans were no longer a priority for the new administration and all plans were frozen. After 5 years of planning and community saving everything had ground to a holt. This caused tensions between Baandhani and Shelter Associates as trust was fading between the partnership. The relationship between Shelter Associates and the new municipal commissioner had deteriorated. The community members were starting to lose patience and faith in Shelter Associates, which further affected their relationship and also led to the creation of fractions within the Baandhani federation (Shelter Associates, 2004).

Shelter Associates decided to do something proactive and began the registration process to establish cooperative housing societies. Mr Ratnakar Gaikwad was the Commissioner of Registrar at the time and was very supportive and helped Shelter Associates and Baandhani to register the residents in groups of 40–45 members by January 2003.

In the same year Valmiki Ambedkar Malin Basti Awas Yojna (VAMBAY) was introduced by the government in Maharashtra. VAMBAY is a major programme for the provision of affordable housing and infrastructure. This programme receives Rs. 50,000 as a grant from the central government. With the introduction of this scheme, the PMC was now encouraged to implement slum-upgrading projects. Slums were now back on the administration's agenda. The notified housing agency for the Government of Maharashtra; Maharashtrian Housing and Area Development Agency (MHADA) (which has the responsibility to build houses for the EWS) placed an advert in the local newspapers inviting tenders for the relocation of slum dwellers to the Hadapsar site that had previously been ear-marked for Kamgar Putala. It emerged (to Shelter Associates) that MHADA had already drawn up plans for the relocation of 800 families onto this plot and were inviting quotations from interested contractors or NGOs to implement the design. The PMC had also already installed the infrastructure on the site needed for roads, drainage and water connections. The KP residents, however, were not satisfied with MHADA's plans due to the small dimensions of the tenements. The previously submitted Shelter Associates housing plans, which had been developed in collaboration with the community, did not seem to have been considered. Shelter Associates did nevertheless submit a bid for the work, but it emerged that the VAMBAY funds would not be sufficient to implement the scheme as per MHADA's specification and so some contractors withdrew (Shelter Associates, 2004).

Then, in March 2003 there was yet another change of administrative head. Mr Sanjay Kumar took over as the Municipal Commissioner of Pune. Shelter Associates and Baandhani again presented the surveys, research and planning they had done since 1997 for Kamgar Putala to move to Hadapsar. Mr Kumar was sufficiently impressed and decided to award the rehabilitation project to Shelter Associates. The VAMBAY committee also allowed Shelter Associates to re-design the houses and submit designs that suited the aspirations of the community while fitting in with the existing Hadapsar site infrastructure. The Buildings Department of the PMC cooperated with Shelter Associates and supported them to comply with Floor Space Index and other rules and regulations.

8. House design and cost

The 176 house tenements implemented by Shelter Associates were organized in clusters of eight tenements around small courtyards. Shelter Associates had demonstrated that low-rise, high-density solutions work well for the poor, as often their homes double up as workshops and businesses for income generation. Shelter Associates states that slum dwellers rehabilitated into high-rise buildings often sell and move back to the slum as this building style fails to satisfy the needs of the poor to provide space for income generation activities; that they are difficult to maintain over time and that with weak service networks it becomes particularly difficult for families on the upper floors (Shelter Associates, 2004) (Figure 4).

It is for these reasons that Shelter Associates' architects designed the new housing at Hadapsar in low-rise (two-storey) modules of eight tenements, four on the ground floor, and four on the first floor. The tenements are clustered around a central courtyard with one flight of stairs leading to the first floor. The central courtyard is flanked asymmetrically by two small courtyards that provide access to the individual tenements on the ground floor as well as to the staircase leading to the first floor. This design allows plenty of sunlight and ventilation into each tenement. The placement of people's entrance doors were spaced to avoid disrupting circulation in the courtyards and corridors. The design process was done very much in consultation with the residents whose considerations were prioritized and balanced with design constraints. The community chose to have one large window and five concrete *jailis* (trellised openings). The community also took a vote on the option of communal toilets to maximize space in their houses or for the option of individual private toilets. The majority preferred to have a private toilet and so this was incorporated into the design. The internal floor area of each tenement is 200 ft^2 made up of two 10 × 10 ft bays positioned linearly. The tenement designs include a toilet, a *mori* with a kitchen area and a multipurpose living/dining/sleeping area in the 'front' room/bay.

A very clever aspect of the house tenement designs was a raised ceiling height to 14 ft in one of the bays. This allowed for the construction of a mezzanine floor of approximately 100 ft^2 which could be added as and when the family felt they needed the extra space. This effectively gave each household a tenement of 300 ft^2.

The tenement size was an issue for Shelter Associates. According to government rules, families were entitled up to 22.5 m^2 of built up space. However, the VAMBAY funds that had been allocated to the project were not sufficient to construct the quantity of tenements needed at this size, and therefore, the size of the units had to be compromised to fit the budget (Figure 5).

9. Funding and external support

Earlier on in the relocation planning Shelter Associates contacted the Housing and Urban Development Corporation Limited (HUDCO),

The layout of 176 tenements showing clusters of eight tenements around small courtyards.

FIGURE 4 Tenement layout (Shelter Associates, 2004).

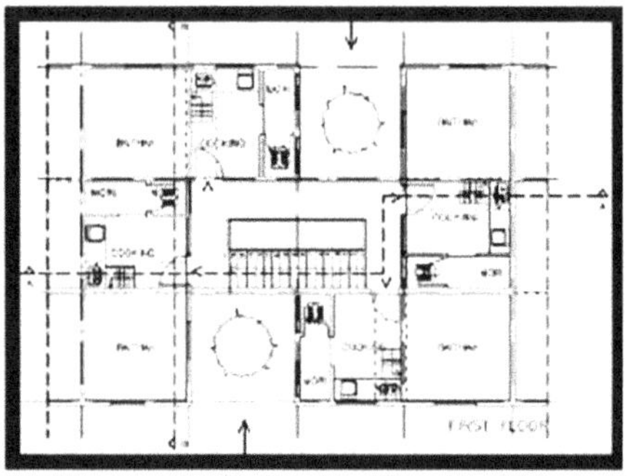

Basic floor plan of the cluster – 4 on GFL and 4 on FFL

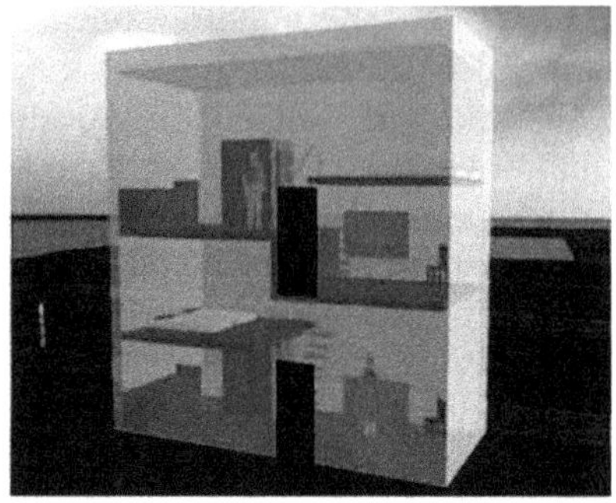

Typical section to show the increase in height in one bay

FIGURE 5 Tenement design (Shelter Associates, 2004).

Mumbai, to begin negotiations for housing loans. The MHADA was also contacted to utilize the government's VAMBAY scheme. This scheme provides a subsidy from central government, routed via HUDCO to MHADA to slum-dwellers in need of housing rehabilitation. VAMBAY is a subsidized scheme with 50 per cent funding from central (Indian) government and 50 per cent from the state (Maharashtrian) government. Along with the scheme the central government set targets to individual Indian states to encourage the utilization of the funds and benefits reaching the poor. The communication between Shelter Associates and MHADA successfully resulted in obtaining a subsidy of Rs. 50,000 to be allocated for each household from VAMBAY; however, this amount later decreased to Rs. 47,619 after imposing various taxes and fees.

The residents of Kamgar Putala's saving efforts since joining the Baandhani federation resulted in each family depositing Rs. 5,000 in a common account opened with the support of Shelter Associates for housing.

The cost of each tenement was initially calculated by Shelter Associates to be approximately Rs. 80,000. Each family had already saved Rs. 5,000. Along with taxes and materials cost rises, an additional Rs. 30,000 needed to be raised.

Shelter Associates set about meeting with banks to organize loans, however Shelter Associates found that banks were reluctant to lend to the poor despite the favourable position of the Kamgar Putala households and the proposed relocation:

- the land was sanctioned by the PMC;
- the drawings had been approved by the PMC;
- there was a subsidy of Rs. 50,000 from VAMBAY;
- the families had saved up Rs. 5,000 each;
- the people were willing to mortgage their tenements;
- the remaining amount needed was less than 40 per cent of the total cost (Shelter Associates, 2004).

Shelter Associates and Baandhani met with several nationalized banks who shared the same concerns of lending to the poor. Some requested that Shelter Associates become guarantors for the loan, but this was not feasible for Shelter Associates. Another of the bank's concerns was if VAMBAY funds were not disbursed. Eventually, the United Western Bank agreed to lend to families with members who were employed within the formal sector. This resulted in 11 of the households obtaining loans. Later, a successful meeting and further meetings and presentations with the manager of The Bank of Maharashtra, Shivajinagar branch eventually resulted in another 69 loans being sanctioned. The strategy was for there to be modules of three families, where each family became a guarantor for the other two. This enabled the families to mortgage their tenements to the bank and receive loans at 8.5 per cent interest, repayable over 7 years with repayments of Rs. 486 per month which was agreeable to the households. The residents had already discussed the amount they would be able to repay without excluding the less well-off. Usually, most families had to spend Rs. 3,000 per year to maintain, replace tin sheets and repair their houses after the monsoon rains; an amount they would now be able to save. Even those families who were tenants at Kamgar Putala were able to save money as they no longer needed to pay the high rents at Kamgar Putala's city centre location.

> The loan issue was very critical to the success of this project. It brought out the dire need for setting up funds from which the poor could borrow. The Bankers were partly justified in their fear that there would be defaults. Their previous experiences with the poor had not been very encouraging. Though the tenements were mortgaged, the bankers were apprehensive of political interference, which would prevent them from taking over tenements of willful defaulters. But in order to make housing a vehicle for transformation, it was necessary that the financial system get institutionalised. Most bankers suggested that the PMC should become guarantors for these loans, as it would be easier to tackle willful defaulters with them (Shelter Associates, 2004).

10. NGO and government partnership

Construction of the houses commenced on site in September 2003 soon after *bhumi poojan*, the groundbreaking ceremony. As MHADA was the nodal agency for the disbursement of funds, their chief engineer was appointed to be the District Chief Housing Officer and responsible for monitoring the construction work. Shelter Associates and MHADA drew up mutually agreed terms and conditions for the release of payment for construction.

The Municipal Commissioner chaired VAMBAY committee meetings which were made up of officers from MHADA, the PMC and elected representatives. This committee agreed to meet monthly to review the progress of any VAMBAY projects ongoing in the city, but in reality 'very few' meetings actually happened (Shelter Associates, 2004). A lack of coordination and effective communication between the PMC and MHADA became a problem for Shelter Associates and significantly impeded the progress of the Hadapsar project. At some points payments were blocked from reaching the NGO for the construction, despite the terms laid out in the original agreement between the parties. 'At one point the PMC decided to levy a premium on land for beneficiaries, which was worked out at Rs. 1,500 per m^2, and recover this from the final bill of the NGO Shelter Associates' (Shelter Associates, 2004). This issue was taken up with the Housing Secretary as the poor were not capable of raising this extra fee for the land and it had not been a precondition in the previous agreements. The Housing Secretary agreed with Shelter Associates and directed the VAMBAY committee to make the land available free of cost to the beneficiaries. However, the PMC is yet to formulate its agreement with the beneficiaries regarding land (Shelter Associates, 2004).

As Shelter Associates is an NGO and not a building contractor – although they were performing the duties of a contractor for this project – there were times that they encountered hurdles during the project planning. One major issue was organizing the electricity connection to the site during the construction process for which neither the PMC nor the electricity board were taking responsibility (Shelter Associates, 2004). Although in charge of the construction process as a building contractor would normally be, the NGO did not have the funds in place to pay for the construction process. This required a large amount of discussion with the PMC and MHADA in order to organize for a 10 per cent advance to pay for one bay of 24 tenements. The release of funds required much paperwork, the submission of consent letters from the validated beneficiaries, on stamped paper, in a format prescribed by MHADA (Shelter Associates, 2004).

Shelter Associates later realized that the Rs. 50,000 subsidy that had been allocated for each tenant was in fact Rs. 47,619 as MHADA was retaining 5 per cent as monitoring charges. Other charges included work contract tax at 4 per cent, and security deposits were another 4 per cent. This reduced the grant available to each tenant to Rs. 43,500 (figures quoted from Shelter Associates documentation. Reason for discrepancies unknown). Shelter Associates raised this issue with the Housing Secretary, the Housing Minister, the Municipal Commissioner and also the Vice President of MHADA; all agreed that the works contract tax was unnecessary and instructed the Commissioner of Sales Tax to waive this 4 per cent fee. After Shelter Associates' requests, MHADA agreed to waive their 5 per cent monitoring charges. However, Shelter Associates were only ever able to claim Rs. 47,619 despite the earlier agreements.

During the progress of the project the cost of steel and cement prices rose, the beneficiaries were forced to cover the cost discrepancies by increasing their loans to Rs. 30,000.

11. Moving

Some families started to move to their new homes at Hadapsar in June 2004. The community decided to allocate houses by randomly drawing lots. When families knew which tenement was theirs they became even more actively involved in finalizing the construction. Many people spent their spare time on the site supervising the work to ensure good quality craftsmanship. The committee also decided to assign the task of fabricating the doors and windows to one of the community members who had a small fabrication unit. This helped to ensure the quality of the work. When the basic shells of the tenements were complete, many families started to further fit out and install flooring, kitchen sinks, work surfaces and tiling into their new homes. The security of tenure gave them the confidence to make further investments in their homes. A 'stand up kitchen' was one of the most popular installations the households fitted. 'Their middle-class aspirations were being expressed in the use of space and finishes that they chose' (Shelter Associates, 2004).

In August 2004, the floods hit Kamgar Putala badly. Much of the original slum went under water again and many families had to be shifted to local schools. However, by this stage nearly 100 tenements were complete and many of the flood-affected families were able to move directly into their new homes. The change that was taking place was tangible. There was a sense of pride amongst those who had shifted to the new site. Now they were staying in a society and not a slum (Shelter Associates, 2004).

12. Revisited in 2010

There were obstacles in planning the resettlement of the residents of Kamgar Putala to new homes on safer land in Hadapsar. The process was long but eventually resulted in 152 families to be relocated to safer homes since 2004 (Joshi, 2010). The upgrading project was very much driven by a collaborative NGO and CBO partnership which enabled the beneficiaries to be in control of their housing situation.

This housing project has been re-visited in April 2010, nearly 6 years after the residents moved in, in order to assess the present situation and sustainability of the resettlement upgrading programme. Interviews and group discussions have been conducted with key stakeholders to gather their perceptions of the project now.

Key issues that emerged from the interviews include:

Sense of ownership: Not one resident was found to have moved from their new home, sold up and moved elsewhere. Some residents are still awaiting their ownership documentation from the government, which could block the option of moving, however no residents expressed such a desire. There was a significant sense of pride among residents who feel connected to their homes both financially and spiritually. This has been demonstrated by significant investments in their homes, the installation of mezzanine floors, decoration, fixtures and fittings and gardens.

Community cohesion: There are strong social networks in the community and active CBOs which are enforcing the sense of society. The building design that allows separate living and sleeping spaces was reported to have improved children's concentration on educational studies and enables privacy. The layout of low-rise tenements with a courtyard arrangement has not disturbed community cohesion as can often occur in high-rise slum rehabilitation schemes.

Structural integrity: Residents are very positive about now living in well constructed, concrete buildings rather than temporary shelter. The advantages of living in a water-tight environment which does not allow the entry of snakes and insects has made residents feel more secure as well as had a significant impact on improved health.

Maintenance: After 5 years, some of the external plaster work is beginning to show wear which some residents are unhappy with. Due to tight budgets and the NGO being forced to coordinate construction (which is not normally their area), the standard of finish was compromised. Residents are now responsible for this maintenance but existing CBOs and community savings groups are well established to deal with such issues.

Water and sanitation: Residents report a main advantage of the new housing is the provision of a private toilet and washing area within their home. Compared to the use of communal toilet blocks in their previous slum, they no longer have to worry about security issues when using the toilet at night. Residents also report a lower incidence of waterborne diseases since resettling. Water supply to the Hadapsar site is not 24 h as was the norm at Kamgar Putala; however, residents did not report this to be an issue. The desire to have taps and private water connections into the tenements (rather than the present communal stand-posts) was expressed and residents are aware that they are able to invest in installing the pipe-work themselves should they wish. The infrastructure for Hadapsar was provided separately by the government authorities. The NGO's house designs were laid out to fit around the existing site infrastructure.

12.1. Distance from amenities

Although Hadapsar is 8 km away from Kamgar Putala, the residents chose the location themselves from one of the sites reserved for the 'EWS' that were offered to them by the government. Some residents have reported the displacement to the outskirts of the city initially caused difficulties with distance to bus stops and therefore transport to schools, markets and job opportunities. However, residents now report that a new bus stop has been created and alleviated the situation.

12.2. Issues

The residents are generally highly satisfied with their new homes; however, some reported they would like external street lighting, water supply and better plasterwork. The first two issues are the responsibility of the government whose remit is to provide the site infrastructure. The aging buildings after only 5 years is the most significant issue residents are disappointed with, and which is due to a very tight construction budget. The NGO and CBO partnership utilized the funds available to the best of their ability and made the decision to organize the construction of the shell of the housing themselves in order to save costs. Residents agreed to incrementally finish and fit out their homes with time. The partnership did the best they could considering the financial circumstances and have still achieved a significant improvement in quality of life of the residents.

13. National slum policies

The development of Hadapsar demonstrates a successful case of community-led processes in risk reduction and participatory planning. However, since Hadapsar, various slum-upgrading policies now exist in Maharashtra. India's national slum policies introduced in the last 10 years now take into more consideration the involvement of communities.

In 2001, The Draft National Slum Policy was formulated by the government and developed a slum resettlement plan for 60,000 people. 'This initiated a newer level of understanding of the issue of urban slums; that slums are an integral part of urban areas and contribute significantly to their economy both through their labour market contributions and informal production activities'. Over the years the policy in India has redefined slums to include all under serviced areas. It does not advocate slum clearance schemes except under strict resettlement and rehabilitation guidelines which are defined in respect of slums located on untenable sites (Asha Seattle, 2006).

The Mumbai Urban Transport Project required the resettlement of approximately 60,000 people living close to train tracks, facilitated with popular participation and partnerships with NGOs and the National Slum Dwellers Federation (NSDF) were involved with the policy formulation. This project demonstrates the importance of partnerships between stakeholders, donors, state and local bodies, CBOs and NGOs (Risbud, 2003).

In August 2001, the Prime Minister announced a major programme called VAMBAY for the provision of affordable housing and infrastructure. This programme receives 50 per cent subsides and 50 per cent loan components with a grant of US$200 million from the central government. It was under the VAMBAY scheme that the Hadapsar resettlement project managed to raise much of its funds.

In December 2005 the Jawaharlal Nehru National Urban Renewal Mission (JNNURM) programme was established with an emphasis on strengthening infrastructure and housing for the urban poor. Pune city was a recipient of funding under the JNNURM (BSUP) and identified 40,000 families currently residing in non buildable zone like river beds, hill top/slopes etc. in need of relocation. This accounts for roughly 20 per cent of the total slum households in the city. Pune city has been the largest beneficiary of NURM funding under the BSUP programme to relocate affected families. The plans already authorized in Pune total over Rs. 640 crore (JNNURM, 2007; Shelter Associates, 2007). 'Public policy has seen a shift from the role of controller and provider of housing to facilitator of housing. Various policy decisions have influenced housing supply in the city, which has had a bearing on the growth of slums in the city' (Risbud, 2003).

As well as VAMBAY and JNNURM other slum improvement schemes overseen by Maharashtra Housing and Area Development Authority (MHADA) include Lok Awas Yojana, Integrated Housing and slum development programme, National Slum Development Programme and the Khandesh Vikas Package Programme (MHADA, 2010).

The Slum Rehabilitation Authority has another model for slum rehabilitation which creates incentives for private developers to take on slum rehabilitation by granting Transferable Development Rights which allows the sale of additional built-up space for profit. The SRA brought about a new policy which recognizes the rights of any slum and pavement dwellers resident in the city since 1 January 1995 to 'avail of an alternate permanent accommodation' (Homeless International, 2010). Under the scheme the SRA provides 100 per cent free housing to eligible families in the form of a tenement with a floor space of 269 ft^2 (approximately 25 m^2) (Kolte, 2010). The policy incentivizes private developers to rehabilitate slum-dwellers in new buildings in situ, with the exception of slums located on non-buildable or dangerous zones of the city such as hill tops, slopes, in the vicinity of rivers and *nallahs*.

14. Final remarks

The residents of Hadapsar are well aware of the other slum upgrading programmes in Pune city which provide larger tenements for free to slum dwellers that are rehabilitated from settlements on more valuable land. Due to the hazardous nature of the flood-affected site upon which Kamgar Putala was located, the residents mobilized with the assistance of Shelter Associates utilized the government schemes they were eligible for help under, and achieved their goal of safe housing away from the hazards of the water. As well as moving away from the threat of floods, the residents were able to use the opportunity to 'build back better' and have created permanent housing considerably better than their previous slum home.

Resettlement schemes can often fail due to the distance created between a community's place of origin; however, the clearest indication of the success of Hadapsar is that no residents were found to have sold up and moved out, even nearly 6 years after habitation. The success of the scheme can be attributed to the housing delivery model using a community partnership which embedded sustainability principles throughout the conception, organization and implementation of the scheme. As the recipient end-users were fully included in every step, an unshakeable sense of ownership and connection of the community to the new development has been created.

The residents at Hadapsar are no longer threatened by the annual devastation of floods and most stakeholders would say the resettlement project has been a great success. After inhabiting the new houses for 5 years, many residents are now used to their new situation and have increased expectations for their future quality of life.

The community consultation and participation organized by the NGO has been exemplary following the guidelines that many slum-upgrading agencies advocate. The small Indian NGO worked well with communities with the use of a small team of social workers who were well regarded by the community and therefore able to form a strong partnership for implementing the resettlement project.

The success of this resettlement project has positively affected an entire community and the lessons learned have been promoted both nationally and internationally within best practice documentation for slum upgrading which has had a beneficial impact on strategies for slum upgrading in India.

Residents have invested significantly in decorating and fitting out their new homes, some have installed piped water connections into their houses, some have planted gardens and decorated their interiors; these all indicate a sense of pride of their home. One resident has established a shop within her home (Figures 6–8).

After assimilating the data gathered during stakeholder interviews and group discussions the key issues that repeatedly arise are the detrimental effect that the resettlement has had on residents' water supply and access to livelihoods. The residents are now no longer affected by the threat of floods, but are now worse off since resettling in these respects. Access to water and livelihoods are basic human needs which residents were not aware that they would have to sacrifice in exchange for a less vulnerable house structure. Residents are not content with the current situation and many have expressed a desire to move in order to gain these basic needs.

Due to the limitations of MHADA grant funding the construction standard of the houses was highly compromised requiring significant investment from residents which has made the move not cost effective for many residents particularly now that their livelihoods are at detriment since resettling. Infrastructure deficiencies are the responsibility of the PMC and are particularly affecting the residents' perception of success of the project.

FIGURE 6 Decorated and well-equipped houses.

FIGURE 7 Self-made water connections and gardens.

The resettlement project lacked the creation of employment generation activities. No business units were provided or community facilities such as meeting halls, children's playgrounds, etc. These are all vital components of a community which are significantly lacking at Hadapsar. The failure to include these has had major consequences on the residents' livelihoods. However, in 2010 Shelter Associates have started to work with residents to promote livelihoods generation activities through sari recycling.

FIGURE 8 Home-based businesses.

Since this resettlement project was designed under the MHADA grant, policies for slum upgrading have moved on and the stipulated conditions developers are required to provide to rehabilitated slum-dwellers are of a higher standard. The residents at Hadapsar are no longer a priority to the NGO and PMC who now consider this project complete, however the residents are experiencing problems which are the responsibility of the PMC to fix. The stakeholder's ignorance of the community is poor and shows that the long-term welfare of this section of society is not of priority. Handing over and walking away is inappropriate, irresponsible and against sustainability principles not ensuring a plan for the future management and maintenance and future scenarios for the development. However, the NGO social workers do receive communication from the residents and have at times supported residents by accompanying them to local government offices when lobbying the PMC for action.

Due to the vulnerable site conditions at Kamgar Putala residents on the banks of the river needed to be relocated, but the effect of the distance of separation from the site at Kamgar Putala has had serious consequences on the residents' quality of life due to the impact on livelihoods. Since the community have moved to Hadapsar, the PMC have now constructed a flood-retaining wall which now protects the rest of the residents who stayed at Kamgar Putala. These residents are now safe and *could* have better prospects than those who relocated to Hadapsar because they are now going to be included in a new government Slum Rehabilitation Authority (SRA) slum-upgrading scheme. This means they will be the recipients of a new house, for free, with a private toilet, bathroom and bedroom, within high-rise buildings on the same site at Kamgar Putala. Therefore these residents will not be required to move from their city centre location within easy reach of their existing livelihoods, they will not be required to contribute financially to the cost of a new house, and the new house will provide more space, better facilities and construction. However, many residents included within SRA schemes have complained about inappropriate building designs and poorer quality of life after rehabilitation. Also, SRA projects are notoriously slow to take off, this means that while the residents at Kamgar Putala are still residing in slum conditions, their old neighbours at Hadapsar are benefitting from a healthier living environment.

However, the residents at Hadapsar are benefitting greatly from the low-rise courtyard building design with plenty of open space around which they would not have had if stayed at Kamgar Putala. This effective building design has been particularly successful as it made use of the NGO's architects who are experienced in the impact of building design on social cohesion and living styles of the urban poor. The buildings were also designed in consultation with the residents and have a very good impact on residents' lives. Considering the circumstances the NGO did the best they could with the resources available for the residents. The PMC could learn greatly from the participatory techniques and community relationships effectively fostered by the NGO during the development of this project.

In situ resettlement would have been ideal for the flood-affected residents at Kamgar Putala. Water supply is a basic human need that should have been prioritized along with other infrastructure by the PMC. If comprehensive planning strategies were in place for the city of Pune, the residents of Kamgar Putala who were relocated could have been spared the upheaval of moving and instead included in the in situ SRA scheme after the construction of the flood-retaining wall. However, the time scales involved would have meant a poorer quality of life for the residents in the medium term, therefore for most residents at Hadapsar, relocation was the best option and has spared the upset of devastating floods for the past 5 years. The problem of social housing is very complex and hence cities need to work out strategies- both long and short term based on sound information systems which could explore combination of government and public private partnership models keeping in mind that scarce resources like land should be optimally used.

Now that many residents are indicating a desire to move once they receive their ownership papers, this may indicate a struggling sense of community ownership and a failing in sustainability of the resettlement project. However as residents have had secure tenure for over 5 years, upward economic mobility might be the reason that people are aspiring for bigger and better homes, particularly in Pune, where poverty can be partly attributable to a lack of access to secure tenure. The residents have now been living at Hadapsar for over 5 years during which time their expectations have raised and, as is human nature, their desire to improve themselves further is ever present. The community s resettlement to Hadapsar has been a successful step-up for the community which has encouraged the residents to want to better themselves and hold higher aspirations for their future generations.

> We like the present house because we are free from the recurring problem of floods. We had to bear lot of problems previously, like lot of mosquitoes, the problem of cleanliness, which we do not face now. We used to find snakes in the house after the flood water receded from our house, the house used to stink. We were facing these issues every year ... We had to spend quite a lot each year on the maintenance of the previous house due to the floods. We had to almost rebuild our house every year after the floods. This [house] is better than that one and bigger (Patel, 2010).

> We used to stay in a tin shed. Now it is a concrete house. Children have a good school nearby. Good road is nearby. Everything has become easier.... Our life has improved here. The house is big enough to accommodate all of us. Children can study well. We are living comfortably and have no tension regarding this house (Kamble, 2010).

> After coming here our lives have changed. Children have a space to play, we also have more open space to use, the house is big, the toilets are inside the house. The lifestyle has improved after coming here (Pathan, 2010).

Acknowledgements

This paper draws from a PhD investigation of the sustainability of slum-upgrading approaches through stakeholder perception in Kenya and India. Detailed analysis of the role of key stakeholders and sustainability evaluations has not been included here, but could be the subject of a future paper. For further information regarding this research please contact the author. The residents of Hadapsar and Kamgar Putala. Pratima Joshi, Shelter Associates. Sandyha Kamble and Shashikant Asware, Social workers, Shelter Associates.

Glossary of terms

Boudh, Buddhist;
Crore, 10 million;
Jailis, Trellised openings;
Kutcha, Non permanent, lower quality, dried brick, mud or tin;
Nallah, Stream;
Patra, Tin sheets;
Pucca, Permanent, higher quality, brick, concrete;
£1 = Rs. 73 (correct as of May 2011);

In this report the terms slum and informal settlement are used interchangeably.

References

Asha Seattle, 2006. Urban slums in India – the myths and the reality. *Asha Kiran* 12(2). 1–3

Homeless International, 2010. *The Slum Rehabilitation Authority and Transferable Development Rights*. www.homeless-international.org/standard_1.aspx?id=0:37913&id=1:32596&id=0:380 &id=0:277&id=0:262 (accessed 27 August 2010).

JNNURM, 2007. Jawaharlal Nehru National Urban Renewal Mission, Ministry of Urban Development, Government of India. www.jnnurm.nic.in (accessed 1 February 2011).

Joshi, P., 2010. Shelter Associates Director Stakeholder Interview, Pune.

Kamble, B. 2010. Hadapsar Resident Interview.

Kolte, S. 2010. Interview. Additonal Chief Executive Officer, At the SRA Office, Pune.

MHADA, 2010. *MHADA Nashik Board Activities*. http://mhada.maharashtra.gov.in/?q=board_nashik_Activities (accessed 14 August 2010).

OCHA/IRIN and UN-Habitat, 2007. *Tomorrow's Crises Today: The Humanitarian Impact of Urbanisation*. AusAID, UN-HABITAT, IRIN. ISBN 978-92-113-1964-4.

Patel, B., 2010. Hadapsar Resident Interview.

Pathan, R.M., 2010. Hadapsar Resident Interview.

Risbud, N., 2003. The case of Mumbai. Understanding Slums: Case Studies for the Global Report for Human Settlements. 15. Development Planning Unit Projects, University College London. www.ucl.ac.uk/dpu-projects/Global_Report/pdfs/Mumbai.pdf (accessed 23 June 2011).

Sen, S., Hobson, J. and Joshi, P. 2003. The Pune slum census: creating a socio-economic and spatial information base on a GIS for integrated and inclusive city development. *Habitat International* 27(4). 595–611.

Shelter Associates, 2001a. *Baandhani at Kamgar Putala*. Shelter Associates, Pune, India.

Shelter Associates, 2001b. *Kamgar Putala: Baandhani Women Welcome the Media*. Shelter Associates, Pune, India.

Shelter Associates, 2004. *The Kamgar Putla Story: A Community's Struggle from a Slum to a Society*. Shelter Associates, Pune, India.

Shelter Associates, 2007. *Study on Comprehensive Rehabilitation Strategies for the Urban Poor in Pune City*. Shelter Associates, Pune, India.

SPARC & Shelter Associates, 1998. *By The Water*. Shelter Associates, Pune, India.

United Nations Department of Economic and Social Affairs, 2006. *World Urbanization Prospects*, The 2005 edn. United Nations. http://www.un.org/esa/population/publications/WUP2005/2005WUP Highlights_Final_Report.pdf (accessed 1 February 2011).

United Nations Human Settlements Programme, 2003. *The Challenge of Slums*. Global Report on Human Settlements. UN-HABITAT, Earthscan.

United Nations Population Fund, 2007. State of the World Population Report, Unleashing the Potential of Urban Growth, UNFPA. http://www.unfpa.org/swp/swpmain.htm (accessed 1 February 2011).

Lost in translation? The challenges of an equitable post-disaster reconstruction process: Lessons from Chile

CAMILLO BOANO[1] AND MARISOL GARCÍA[2]

[1]UCL, Development Planning Unit, 34 Tavistock Square, London WC1H 9EZ, UK
[2]International Programme, Building and Social Housing Foundation. Coalville, London, UK

Built environment experts, steeped in their default modes of operation, have been critiqued in post-disaster scenarios as silent perpetrators of unsustainable solutions that focus solely on physical deliverables, rendering them as 'useless'. This paper, grounding the analysis in the recent post-earthquake landscape of Chile and its reconstruction plan in Constitución, adopts a theoretical framework of post-colonial discourse to elucidate the significance of the notion of the 'expert as translator', activating the capacity to amplify the voices of those excluded by dominant forms of knowledge production. Such an attempt would aim to challenge architecture, urban design and the built environment – as disciplines and as practices – in planning for equitable reconstruction. Aligned with Lefebvre's socially produced space, the paper seeks to enlighten the idea of encouraging cultural agency and constructive insurgency in a post-disaster context. It examines how the adaptable, multiple, open, integrative and transformative qualities of reconstruction spaces are essential in responding to cultural differences and dynamic processes.

1. Introduction: The shift to interpretive agent

Reconstructing homes and infrastructure and rebuilding communities, livelihoods and entire cities devastated by disasters is a complex task. This complexity emerges and is magnified in political and institutional dimensions where various actors with often conflictive agendas, powers, visions and mandates are involved in a dense and constricted environment characterized by the primacy of speed, performance and numbers. Spatially there are massive needs for the physical reconstruction of damaged structures, provision of new housing for affected populations and the conservation of heritage buildings which encourage the bad practice of only issuing deliverables and not introducing opportunities to incorporate preventive measures aimed at reducing the frequency and severity of future disasters. All these complexities are nested in a societal arena where restitutions, compensations and resettlement, as well as the erosion of livelihoods and security, potentially prevent social cohesion, thereby undermining transformative development (Boano, 2009).

Building on this, central to the paper is the conception of contemporary post-disaster intervention as located within a web of contested visions, rendering the production of space as an inherently conflictive process where competing forces and contentious actions both for and against the status quo are manifested. This contested space calls forth what Appadurai (2003) termed the 'illusion of permanence' – a phrase that was originally the title of Francis Hutchins 1967 book (*The Illusion of Permanence: British Imperialism in India*). This resonates because it captures a kind of desire of the imperial system, while simultaneously presenting the anguish and ambivalence involved in such production and representation: the arrogant conceit of certain grand projects, but also the humble yearnings that ordinary people seek constantly in their everyday lives. What intrigues us most in that vision is not only the continuous tensions and disparity between grand schemes and grounded experience, but the strategies, inventions and narratives that produce stability, an imagination of tomorrow and potential alternatives. In other words, we emphasize the different 'readings' that experts have in imagining and producing new post-disaster recovery plans and the relation of those interpretations with everyday life.

Such readings are often done by design professionals, development practitioners, architects and shelter managers through working, imagining, producing and representing not only housing solutions but also images and visions of a post-disaster recovery. Their role as experts is extensively discussed by Lizarralde et al. (2010), suggesting that the contribution of architects in a post-disaster situation comes from *an appropriate interpretation of stakeholders' varying capacities* rather than through their traditional roles in design and construction. Zetter et al. also question their role by defining the action of place-making as 'not just the domain of the powerful or of design professionals' (2010, p. 210) but as 'a process where ordinary people create a bounded, identified, meaningful, named and significant place' (p. 210). Stating that architects are the last people needed in disaster reconstruction, Sanderson (2010) provoked an interesting debate around their role, their relevance and the appropriateness of their knowledge in dealing with the production of space in a post-disaster context.

However, spatial practices such as architecture and urban design, rather than being 'useless', must have a significant influence in the reshaping of space in the aftermath of a disaster. Since its English publication in 1991, Lefebvre's *The Production of Space* has had an extensive influence over the way in which the knowledge, the experience and the practice of architecture have come to be understood, especially in his critique of the mainstreamed perception of space as fixed, immovable and passive. The effects of spatial transformation produced by a disaster manifested as destruction go beyond the built form, encompassing complex overlapping social and spatial issues (Boano and Hunter, 2010). When Lefebvre states, '(Social) space is a (social) product' (1991, p. 26), he considers space beyond its intrinsic physicality, exploring its philosophical meaning in conjugating real and mental space. He defines space in a triad – spatial practice (perceived space), representations of space (conceived space) and representational spaces (lived spaces) – understood by its interconnections (1991). One of the relevant contributions of Lefebvre's work is the conception that space is 'produced' not by a single person but rather 'through a complex set of overlapping societal agencies: the representational, the economical, the phenomenological, the conceptual, the spatial practice of the individual [and] the collective practices of the political' (Till, 2009, p. 126). Lefebvre discusses the role of the architect in the production of space. For him the architect places himself in his own space and has a 'representation of this space' (1991, p. 361); in a 'subjective' space that is limited to graphical representations. He criticizes the idea that 'objective reality' can be attained by those methods of representation (p. 361). This abstract or 'conceived space' is meant to be 'true'; that is 'any plan, to merit consideration, must be quantifiable, profitable, communicable and realistic' (Lefebvre, 1997, p. 144). However, he claims that planned or 'idealistic' spaces are dominated by other 'mode of production' forces, which for the French author at that time was capitalism (1991, p. 360). The 'expert' either works for himself or responds to the interests of 'bureaucratic, financial, or political forces' (p. 365). Space therefore is the manifested expression of conflictive socio-political interests. Who produces space and how does its production affect reconstruction after disasters? Till acknowledging Soja's response to this question states: 'we all produce space' (2009, p. 126).

Once our common responsibility in the complex production of post-disaster space is established, what is advocated is the need for a redefinition – a de-colonization – of post-disaster housing design. What seems to emerge from the adoption of a post-colonial framework and the Lefebvrian production of space is a potentially equitable and inclusive process where built environment experts become 'translators' of the complex set of roles, capacities and powers of the multiple actors involved in the process. This involves a multiplicity of competing socio-spatial requirements where design process is simultaneously the production of physical form, the creation of social, cultural and symbolic resources, and also, critically, the outcome of a facilitative process, in which enablement, resistance and alternatives become central ideas.

In addition, an equitable reconstruction process stresses the need for an ontological rethinking of intervention mechanisms where inclusion, equilibrium and critical awareness need to be sought between providing a development alternative or a form of alternative development (McFarlane, 2004, p. 910), with clear consciousness of either working within the system or seeking to radically restructure social, political and economic relations in the city and beyond. In this role, architecture and urban design have a potential part to play if the latent ability to 'translate' is manifested. Thus, conceptualizing translation in a post-disaster process is relevant for considering both the role of the subject who is 'reading' and the one who is 'writing', as well as for comprehending that this process of transformation is in constant evolution. As a translated version, the post-disaster field should be open to a constant process of re-definition (or re-shaping) of the 'original'. This raises a paramount question: who is the translator in a post-disaster context?

Elaborating on the potential of the disaster expert as translator, and the potential positive agency of translation in the complex redefinition of spatial meanings, this paper adopts a post-colonial theoretical framework to illustrate a case study in Chile and the relevant role of urban design and architecture in the production of space in a disaster aftermath. It aims to adopt an alternative, possibly innovative way of looking at experts as *translators*, aiming to include the voices of those excluded by dominant forms of knowledge production while challenging the spatial representation of reconstruction outcomes and their representations. This is not to say that architects should withdraw from the reconstruction processes, following Sanderson, but rather that their responsibility is shifted to that of 'interpretive agent' (Till, 2009, p. 164) as translator and that the practice and product of such a multifaceted discipline requires one to become 'dependent on a revised version of professional values' (Till, 2009 p. 164). The role of the architect in a post-disaster context is to 'interpret the ways of living of affected residents and the housing typologies of disaster-affected areas, to analyse those ways of living and those typologies and to translate them into technical, organizational and design solutions capable of promoting long-term development' (Lizarralde et al., 2010, p. 24).

The paper first addresses current debates around the post-disaster 'field' of studies, narrowing down the discourse to reconstruction and housing. Afterwards it expands on the notions of translation and representation borrowed from the post-colonial discourse, building an argument for the applicability of these concepts into post-disaster debates. Finally, it presents the Chilean earthquake and tsunami of 2010, examining specifically Constitución and its process of reconstruction.

2. Current debates on post-disaster and the risky approaches of praxis

As anticipated previously, post-disaster recovery is a complex matter. It is not only a process of rebuilding damaged structures, but a process of re-establishing and rebuilding communities, livelihoods, infrastructure, housing and so on, which often lasts for years (Jha et al., 2010). It is 'profoundly developmental in nature, risky because it deals with transformations and equality, and multiple because it takes place across different domains' (Boano and Hunter, 2010, p. 1).

However, the level of complexity of the disaster depends on its scale, nature and corresponding response (Jha et al., 2010), which can result in either the creation of a new disaster (Boano, 2009) or in a process of 'building back better' (Lyons et al., 2010).

Recent events like those in Haiti, Chile, China and Japan highlight the relevance of urbanism and urbanity. Cities could be the safest place in the world when a disaster strikes, or conversely, the most dangerous ones. Effective disaster preparedness, equitable land use planning, appropriate building codes and regulatory frameworks can lessen the impacts of disaster or avoid them altogether (IFRC, 2010). Although it is not an absolute rule, the role of the 'urban' in a disaster aftermath could be significant in reducing the impact of future disasters. In that respect, housing reconstruction should go beyond the mere production of houses and its implications for an individual household, though responses often focus on houses as entities rather than centre on 'recuperating a sense of domestic and public space and place' (Zetter et al., 2010, p. 206), contesting the view of place as a static concept due to its association with 'character' or 'identity' (2010, p. 3). The contested identity of place, not abstracted from everyday life, is precisely what articulates place with Lefebvre's socially produced space. Therefore, in a conflictive post-disaster environment, the 'urban' acquires significance in rebuilding a 'sense of place'. Aligned with this contested and unfixed definition of place, Zetter et al. underline that 'home both as an entity and as a place is in a constant process of consolidation, transformation and adaptation' (2010, p. 209).

All too often, housing is conceived as a one-dimensional and standardized physical artefact, constructed as a reactive, top-down, technology-driven and end-state product (Boano, 2009). Challenging this dominant approach is a fundamental shift from 'bricks and mortar' solutions – the physical dwelling – towards a broader social and economic dimension of housing. Such an approach to housing reconstruction often oversimplifies the complexity of post-disaster situations (Boano and Hunter, 2010). The 'urban', as a hub of confluence of political, economic and social interests, has the potential to be an instrument for spreading equity, thereby attaining the aforementioned complexity of a post-disaster context. Urban design is a transformative instrument which has the power to be either a risky process of shaping foreign and imposed urban form or a collective process of negotiation, which signifies, reinforces and improves peoples' quality of life. The task that built environment experts, as translators, have in planning an equitable reconstruction process is to consider cultural and social diversity and to include the voices excluded by imposed systems of dominance and power.

3. Translating post-colonial debates into post-disaster reconstruction

The post-colonial critique, focusing attention on entrenched and replicated systems of production, aims 'to include the voices of those excluded by dominant forms of knowing' (Sharp, 2009, p. 107). Adopting a post-colonial theoretical framework and using an illustrative case study, the paper aims to elucidate the contributions of both the 'translator' and the 'reader' in post-disaster settings. The 'expert', as translator, should interpret the needs of the affected population, rather than imposing power and dominance. Therefore, post-colonial theory discloses the relevance of raising the voices of the displaced when promoting their active role in a post-disaster situation. The dominance and power of a colonialist approach neglects the ability of the colonized to act, thus post-colonial theory, as a critique, promotes the idea of cultural agency, which can enlighten post-disaster studies and the complexity of reconstruction practices.

Deconstructing the neologism 'post-colonialism' by splitting it into two elements facilitates an understanding of its contested meaning. Colonialism not only removes the wealth and resources of the colonized, but also occupies their territory (Sharp, 2009). As Blunt and McEwan argue, the 'post' can be both; a 'temporal aftermath' and 'critical aftermath' or critical approach that goes beyond, 'but remain[s] closely influenced by colonialism' (2002, p. 3). So 'post' may not only signify what happens after, but also the subversion of the colonized groups to this imposed system of power. For Jacobs (1996), the value that distinguishes post-colonial studies from colonialist approaches is the post-colonial focus in forms of collective insurgency rather than in dominating discourses (p. 28). For the purpose of this paper, post-colonialism as a 'critical aftermath' (Blunt and McEwan, 2002, p. 3) is disclosed both because of its implicit 'critique to conventional ways of knowing' (Sharp, 2009, p. 110) and its capacity to 'present alternatives to dominant [western] constructs' (p. 5) when pursuing an understanding of the implications of the production of social space and cultural interactions in a post-disaster 'field'. This view has potential spatial connotations in removing oneself from traditional centres of dominance and power, and instead operating from the margins. This critical positioning has the potential to produce a separate, and possibly alternative, point of analysis.

In 'Can the subaltern speak?' Spivak (1999) exposes the problematic of speaking for others by distinguishing two senses of the term representation. Representation as 'speaking for' (p. 28), in the sense of political representativeness, and as re-presentation, in the sense of an adequately new presentation of reality 'as in art or philosophy' (p. 28). The subaltern, as she attempts to name, is 'a person without lines of social mobility' (p. 28), being the oppressed non-western society. The language of the subaltern, as Sharp in reference to this text depicts, 'is translated into the language of science, development or philosophy, dominated by western concepts and western languages' (Sharp, 2009, p. 111). This statement is significant due to its questioning of how experts present themselves as 'external' voices when working in the field, requiring also a 'decentring of themselves as experts' (Sharp, 2009, p. 112). In this way, post-colonial discourse serves to present alternatives to the dominant ways of knowing.

4. Translation, a matter of language

The notion of translation extensively discussed in the work of Benjamin (1968) and Bhabha (2004, 2009), among others, is useful in comprehending the role of design in the post-disaster 'field'. The short but influential essay, 'The task of the translator', developed by Benjamin (1968) questions the role of translation, contrasting the 'original' with the translated version. The translation is different from the original, and the task of the translator should be that of redefining the original by offering a new version, which remains closely tied to yet independent of the original once it is produced (Benjamin, 1968; Hernández, 2010).

Benjamin supports the idea of translation not as a copy of the original but as an ongoing and maturing process that redefines the original. Specifically, within this process lies the space of agency for the built environment professional as 'translator' – this is simultaneously re-empowering and daunting. The translation is intimately connected to the original, acquiring significance in its afterlife (Benjamin, 1968). It is because this constant evolution of language that the translation can never be considered as a whole. As well as language, the translation, represented by the metaphor of the *Tower of Babel* by Derrida, is in a constant process of becoming (Wigley, 1988, p. 672). As language evolves, new interpretations will be produced redefining the original again and again.

Following the above, translation in post-disaster processes becomes relevant in considering the role of the subject who is 'reading' and the one who is 'writing', as well as for understanding that this process of transformation is in constant evolution. As a translated version, the post-disaster field should be open to a constant process of re-definition (or re-shaping) of the original, pre-disaster one. However, who is the translator in a post-disaster context? If urban design and architecture practice are considered as processes of translation, whom do these practices serve? What are the roles of the 'reader' and of the 'translator'? Furthermore, the idea of re-defining the original opens up other sets of questions: What are the different views manifested by different processes of translation? What are the hidden structures of power that guide the process? Translation elucidates the idea of power relations and domination systems implied in a reconstruction process. It offers the potential to 're-script' the embedded, dominant narrative for a more spatially equitable post-disaster practice. As the idea of translation is constructive to disclose the process by which the physical fabric is re-shaped, the idea of representation may reveal the inclusiveness of cultural differences in the formation of design products at the urban as well as the architectural level.

5. Representation as formations

Bhabha argues, '[the] stereotype is not a simplification because it is a false representation of a given reality. It is a simplification because it is an arrested, fixated form of representation that, in denying the play of difference [...], constitutes a problem for the representation of the subject in significations of psychic and social relations' (Bhabha, 2004, p. 107). The representation of the subjects as homogeneous stereotypes removes the importance of their own physical, cultural and social relations. Crucial here is that literature conceives representation not only as *deformation* of reality but also as *formation* articulated by an historical process. The contribution of Lefebvre's work concerning the representation of space is again relevant here: for him the architect places himself in his own space and has a 'representation of this space' (Lefebvre, 1991, p. 361), in a subjective privileged manner that is limited to graphical representations. Hernández, while discussing Bhabha's work, further explains this, arguing: 'the problem arises because the stereotype simplifies the represented subject reducing it to a self-contained, complete, immutable assemblage and, so it ignores the historical process through which such an assemblage has been formed' (2010, p. 32). This abstract (conceived) space is meant to be 'true'; that is to say 'any plan, to merit consideration, must be quantifiable, profitable, communicable and realistic' (Lefebvre, 1997, p. 144) where the expert either works for himself or responds to the interest of 'bureaucratic, financial, or political forces' (p. 365) where the risk of reducing the colonized subject into a self-contained assemblage is two-fold. It is dissociating it from its own history by placing the represented subject as an external and 'recognizable object' (Dovey, 2010), and it is reducing the complexity of the context by implying all subjects are the same (Hernández, 2010), thus failing to disclose the importance of cultural difference. Thus, analysis of post-disaster scenarios by the lens of representation is significant to both; by depicting the importance of cultural differences between seemingly homogenized groups and by evidencing the risk of reducing recovery responses to fixed assemblages, and therefore disregarding the historical process by which the devastated place was formerly shaped. Urban transformation processes should consider the connections between responses rather than work in isolated fragments. Self-contained responses oversimplify the complexity of urban social space (Boano and Hunter, 2010), dismissing the relevance of differences. The elimination of differences can be used as a form of control (Hernández, 2010), as for instance, disregarding the existence of certain groups and thus marginalizing them of any possible response. Responses for re-shaping space should consider the value of cultural processes and the process by which the conflictive place was previously formed. Through such a lens of representation, a set of questions might be answered. What are the impacts of reducing housing to an immutable assemblage; the risks to reduce architecture as object-centric and thus superficial in a complex post-disaster scenario (Boano and Hunter, 2010)?

For Dovey (1999) such risk can be avoided or reduced if architecture and urban design are being understood as spaces and places as inherently caught in the complicities of powers. Such complicity is the condition of design. As she states, as the invention of the future such practices are inherently political. 'Architecture and urban design are highly social arts wherein the task is to link aesthetic imagination to the public interests. Community architecture is in a sense, a tautology. Is there any architecture that exists outside the common interests?' (Dovey, 1999, p. 5) or outside a given social plurality? The practice cannot stand outside the relations of power shaping the theories that address the imagination of a new utopia a new grand social dream. The aim, thus, is to look at how adaptable, multiple, integrative, open and transformative the reconstruction solutions are in order to respond to cultural differences and a dynamic timeline. Extrapolating these thoughts into the case study of Chile's 2010 earthquake allows rethinking the ways in which space is produced. From the theoretical framework elaborated in the previous part of the paper, an interpretative tool emerges for examining the case.

Table 1 is an attempt to synthesize how the main concepts disclosed above – translation and representation – by which critical questions were posed, are associated to the role of the 'urban' and 'urban design' in post-disaster settings. The adoption of a post-colonial theoretical apparatus, as a critique to colonialism, attempts to uncover the potential of cultural agency and of architecture as 'transformative agency' (Till, 2009, p. 164) for a recovery process. The critical questions determine a set of normative criteria and indicators, as a lens to analyse the case. Despite being overlapped and interconnected, these indicators intend to qualify the recovery process to reveal that adaptable, multiple, open, integrative and transformative qualities of reconstruction spaces are essential in responding to cultural differences and a dynamic timeline.

Concisely, this paper argues that a reconstruction process after disaster is equitable when:

- 'Translators' acknowledge the significance of the roles and capacities of the multiple actors involved in a recovery processes, encouraging spaces of negotiation and therefore including the voice of the affected population supporting an open reconstruction process.
- Approaches consider the dynamic spatial and social processes together with multiple, adaptable and integrative responses.

For the analysis that follows, the normative criteria within the overall concepts will guide a critical interpretation of the argument by pooling examples of the grounded reality. Though recognizing that many other possible entry points exist, by selecting this criteria and purposefully defining the lenses through which we look, the framework aims to encourage a reflexive understanding of current realities.

6. Chile 2010 earthquake, an illustrative case study

Chile has historically been affected by earthquakes, tsunamis, landslides and volcanic eruptions –events constantly challenging the country's capacity to cope and recover (Figure 1). The strongest instrumentally recorded earthquake in world history, of 9.5 Richter magnitude, is the Valdivia earthquake of 1960. The city of Tocopilla was also struck in 2007 by a 7.7 Richter magnitude earthquake, which left about 15,000 victims and 2,600 houses completely destroyed (ONEMI, 2010). In 2008, the city of Chaitén was covered by ashes from its volcanic eruption, creating a massive displacement of 5,000 inhabitants to nearby towns and cities. An earthquake of 8.8 Richter magnitude with an epicentre located off-shore near Concepción hit the country on 27 February 2010. Impacting an area with almost 80 per cent of the country's population and significant urban centres (e.g. Santiago, Concepción, Rancagua, Talca, etc.), this event left nearly 500 casualties and 370,000 houses damaged of which 220,000 were seriously damaged (MINVU, 2010a).

Chile is a country of approximately 17 million people (INE, 2010) of which about a third live in

TABLE 1 Analytical framework; critical questions, criteria and indicators (The authors, 2011)

	Critical questions	Criteria	Indicators
Translation	What is the role of the 'reader' and of the 'translator'?	Interpretation and understanding of roles and capacities of the diverse stakeholders	• Extent to which residents influence decisions and actively contribute with the recovery process • Extent to which the architects and experts contribute with the recovery process
	What are the visible and hidden structures of power that guide the process? Who are included and excluded from the reconstruction process?	Equitable influence in decision making process	• Extent to which the affected population is included in the decision-making, creating an open reconstruction process. • Extent to which power is being exercised unevenly excluding some of the decision-making, creating closed or hidden spaces in the reconstruction process.
Representation	To what extent the response address the needs of diverse affected people?	Recognition of cultural differences	• Extent to which responses are multiple and adaptable to respond to the diversity and changeable needs, thus addressing cultural differences • Extent to which responses are unvaried and fixed denying to acknowledge cultural differences
	What are the impacts of reducing housing to self-contained responses; as pieces of the city that stand alone and are isolated?	Interaction within processes	• Extent to which responses are integrative within spatial practices • Extent to which responses stand alone and are fragmented

its capital, Santiago. A strong centralism underlined by the concentration of population in its capital also rules its political and administrative structure, which explains the vertical character of urban and housing regulations. Administratively it is divided into 15 regions – internally subdivided in provinces – commanded by regional administrators and governors appointed by the central government. Although locally each council has democratically elected mayors and councillors, the uneven development within its regions is highly visible. Sierra (2006) argues that both the centralized character of policies and distribution of regional funds are barriers towards the pursued decentralization often heard in current political discourse. The Ministry of Housing and Urbanism (MINVU) is responsible for the national housing policy framework and provides national urban regulation guidelines. Specific urban planning frameworks comprising regulatory instruments (i.e. land use regulation, infrastructure, building codes, etc.) are produced by local councils, and approved by the regional government councils. Currently, the government regulates the production of social housing and provides funding through subsidies to the demand, while the private sector delivers the projects. External bodies (EGIS), which must be registered at each regional government, deliver technical solutions and are meant to provide social services, connecting government funding and regulations for social

FIGURE 1 Disaster's location in Chile.

housing with communities and individuals in need. Home-upgrading for homeowners and both new collective construction projects and used housing purchases for dispossessed people are subsidized. No socially rented property model is supported by the current housing policies; instead a supportive ownership-based model is in operation.

Within this context, Chile's 2010 earthquake is selected as an illustrative case specifically encompassing the recovery process experienced by the city of Constitución. The city, of more than 50,000 inhabitants (INE, 2010), is one of the most affected cities due to its proximity to the epicentre, and due to the tsunami, which later destroyed its coastal areas. Among multiple cities seriously damaged, Constitución was selected since it is the first city in which the private and public sector established a cooperative agreement for planning the reconstruction efforts (MINVU, 2010c). PRES Constitución (Sustainable Reconstruction Plan) is a reconstruction plan proposed initially by *Arauco*, a private forestry company that has a major cellulose production plant in the city. Including a team of national and international experts and consultants (Arup, Elemental, University of Talca, Fundación Chile, etc.), PRES is defined as a 'consortium chaired by the local municipality and formed by MINVU (Ministry of Housing and Urbanism) and Arauco' (PRES, 2010a). The public–private alliance goal is to produce a master plan that will guide the reconstruction of the city and 'allows putting in order and coordinating all the mid and long term initiatives of the diverse sectors' (MINVU, 2010c). The model of a 90-day master plan was copied by alliances in other affected significant urban areas of the country (i.e. Pelluhue Talca, Licantén, etc.). The finance of the 90-day master plans relies on donations and private enterprise funds, which in the case of Constitución amounts to the sum of US$1 million entirely financed by Arauco (Aliaga, 2010). However, the scaling-up and implementation of these initiatives is yet unclear. If these master plans are not connected to national investment priorities, resources distribution and financing mechanisms, they will turn into mere graphical representations (Bresciani Lecannelier, 2010); or using Lefevbre's terminology, to merit consideration, these 'conceived spaces' must be 'quantifiable, profitable, communicable and realistic' (1997, p. 144).

After a year, a few projects have been rebuilt in Constitución as a result of the master plan (i.e. school, commercial units, touristic squares, reforestation for tsunami mitigation), and around 30 projects had secured finance or are in process of obtaining it to be implemented (PRES, 2011). However, housing is still a critical debt. Nothing more than the 18 m^2 wooden emergency shelters (Figure 2), often located in 'temporary' camps, and some other individual solutions, usually in home-owners sites, are visible. Citizens' claims for a permanent solution had risen and an estimated number of 5,000 housing units are missing (Tillería, 2011). The government declared 130,000 housing subsidies to be given over the year at a national level, of which approximately 2,700 (including supply and demand subsidies) correspond to the area of Constitución (Valencia, 2010; MINVU, 2011b). However, many, all over the country, are expecting where and when the right declared by this certificate could be claimed, as many of these remain as uncertain solutions.

Thus, several questions arise, related to the quality of the plans: How multiple, adaptable and flexible are the proposed approaches taken in response to cultural differences? What is the value of the 'translation' process? Do these 'translated versions' interpret the urgency and needs of the 8,000 displaced in Constitución? Who are included and who remain excluded from the reconstruction process? These questions are core for the following analysis, which pursues to critically translate the language of theory into the grounded reality.

FIGURE 2 Emergency shelters (Mediaguas) in 'La Poza' sector.

7. The task of the translator: Roles, capacities and equitable influence

Master plans, planning proposals, consultancy reports, among others, are accessible as interpretations of reality made by experts (PNUD, 2010; PRES, 2010c). According to Lefebvre, there is no 'objective reality' (1991, p. 361) but a set of imaginaries or translated versions by experts, which make this an appealing case to discuss. Such competing 'translations' are a potential pitfall in the complex post-disaster landscape. Which translation rises to the top? How do actor power-relations affect this? What is the task of the translator? Are the architects and experts lost in translation? The understanding of the roles and capacities of both 'readers 'and 'translators', and their desirable equitable influence on the decision-making process are the two main analytical criteria that help to elucidate the case.

In practice, the master plan development is led by the architecture firm ELEMENTAL – the world-famous practice chaired by Alejandro Aravena – who work in collaboration with the aforementioned consultancy institutions and with central and local government officials. Arauco is committed to support with resources and with its experience in dealing with contractors, while the government has committed support with housing subsidies and planning instruments for the implementation of the plan (MINVU, 2010c). Although the approach towards city reconstruction has been emphatically top-down at its core, the participation of civil society in the master plan development has been encouraged since the commencement of Constitución reconstruction planning (PRES, 2010a,b). The affected population supported more than 50 meetings organized, gave feedback and actively responded to the open consultancy process done by the end of June 2010. This public consultancy had two objectives: to decide on the priorities of the inhabitants in terms of public projects and public spaces, and to discern if the citizens support the proposed flood park in the coastal area close to the city centre (PRES, 2010a). About 10 per cent of the city population (i.e. 4,120 votes), corresponding to the 50 per cent of the electorate (PLATAFORMA URBANA, 2010), gave their votes achieving a successful citizenship 'consultancy' process. The PRES group erected an 'open house' in the city centre (Figure 3), to receive anyone who wishes to raise their voice and provide information about the planning process to the community, a physical materialization of an inclusive space. Some community leaders become, at the beginning of the plan, active members of PRES as instructors, motivating others to get involved in planning the reconstruction (Hormazabal, 2010). Although there are hidden structures of power influencing the decision-making process, this space of participation is the first attempt to legitimize a process of design while demonstrating the willingness of making this an inclusive process. However, the process seems to be led by centralized decisions, thus being unable to promote citizens agency and to encourage an effective use of local capacities. As Bresciani Lecannelier (2010) notes, the reconstruction requires decentralized management and local agency promotion, as centralized decision-making and lack of coordination of government actions will translate into unfeasible plans lacking local identity, and moreover in the destruction of the already-weakened social networks. In addition, a year after the events, new 'permanent' housing solutions are still unseen while some public projects (i.e. schools, cultural centre, provisional market centre, sport field) also designed by ELEMENTAL are now being implemented (FNR, 2010; PRES, 2010d). Although the budget for implementing some of the plans has been raised from public and private sources, doubts and frustration due to the slow process are increasing in the community (El Mercurio, 2010). Moreover, as Bresciani states, the estimated funding needed for implementing Constitución's plan reaches US$102 million, which corresponds 'to the 25% of the national budget of the Housing and Urbanism Ministry (MINVU) for urban projects in all over Chile' (2010, p. 152). For viewing results of the plans in the city the main risk is to be caught between the planning and implementation processes by creating fixed plans or closed spaces, as the huge investment placed in these initiatives will thus become a loss of time and money and unmet people's expectations. For instance, if housing demands are not sufficiently addressed by the plan, dissatisfaction and distrust towards the government response will raise promptly. A continuous, open and adaptable process should have to be attained as for bridging the gap between both stages.

FIGURE 3 PRES 'Open House' at the city centre.

The architects and urban designers act as translators, being intermediate players between the current needs expressed by the affected population and the economic and political interests concerned in this development process. As we note in the Introduction, the production of space is dominated by other forces. What was capitalism for Lefebvre, for us are surely economic, financial and managerial drivers. In Constitución, Arauco's interest influences the plan by heading and financing this project. Besides building up the social responsibility of the firm, Arauco's interests in undertaking an active role in the reconstruction of the city may be related to securing its labour force and the value of their land and with gaining recognition to be certified under international standards as a responsible business organization. The relocation of Arauco's cellulose production plant has been a sustained debate over the years mainly due to the environmental pollution effects it has over the city. However, it supplies jobs, standing as the main economic activity of the city. Concerning this debate, architects exposed the community to a set of impact mitigation measures during a meeting held at the open house. The location of the plant was given for granted, and was not the central subject of discussion held with the residents during the meeting. How openly was the discussion held? Why is this issue not included in the public consultation? After the meeting, one community member published in the PRES web forum: 'Very good proposals, but we do not have to stay with just what they propose, but to create collective proposals, such as those that have been raised in other forums, meetings or boards' (Cancino, 2010). However, if the inhabitant's priorities and concerns are considered and the plans are met, Arauco's 'hidden' interests will have a positive impact on the city.

A process of reconstruction to be legitimized needs to consider the views of the multiple stakeholders involved and their equitable influence in decision-making process. The case illustrates the complexity of post-disaster processes by raising the question of the role of multiple stakeholders involved in the disaster aftermath, while disclosing the significance of legitimizing the process of design. Closed decision-making process can only result in social exclusion, oppressing the voices of the displaced. There is a risk in considering non-integrated solutions tackling the recovery process by reductive and disconnected responses. The 'urban' gathers an active and significant role as the 'in-between' space, as the collective space in which multiple interests can be negotiated. Architecture can therefore be either a mere tool of representation or become active as an instrument that promotes agency.

8. The whole and the parts: Cultural differences and interaction

The ideas of representation and of assemblages are used as a means to depict the significance of the role urban design has in post-disaster reconstruction. The analysis will be centred in the recognition of cultural differences to uncover to what extent responses are multiple and open to target the multiple and urgent needs and to what extent they are flexible and adaptable to previous and future conditions. Aligning to the idea of assemblage, the case analysis will focus on the interaction within processes, raising the question about the connection between the whole and the parts.

The Chilean central government responded to the 2010 earthquake through the programme called *Chile unite, reconstruct better*. The leaflet directed to the affected population, which explains the scope or the programmes, proposes 'for each victim, a customise solution' (MINVU, 2010b, p. 1) pointing out: 'for each need MINVU had design a solution. Which is yours?' (p. 2). These solutions were classified within seven measures, such as new collective housing reconstruction or on-site reconstruction for homeowners. The language employed depicts the unrealistic approach of the government due to seven measures focused in housing, an over-simplification of reality and a standardized form of production which could not satisfy the diverse, complex and mutable individual and collective aspirations and realities.

The operational tools for nearly all the solutions proposed are based in the conventional non-disaster housing policies' procurement methods, which support an ownership model through subsidizes to demand and private sector project delivery mechanisms, but are also reduce to them relying in previous conditions. Adopting existing formal methods of housing procurement such as new collective housing projects, existing housing purchase, homeowner upgrading programmes and homeowner on-site construction brings positive gains, as benefits for saving time, diminishing conflictive or legal bureaucracy, achieving an easier coordination by drawing on existing knowledge, harnessing current institutions and strengthening the roles of stakeholders that had been working in the housing sector. However, using this knowledge is not sufficient without a coherent grasp and adaptation to the current situation created by the disaster. In the aftermath, such housing procurement modalities do not respond to all the existing realities in a manner consistent with previous methods. For instance, land speculation or unavailability of land made many areas unfeasible for the new collective housing programme to operate. In addition, unavailability of houses for sale due to their destruction excluded many land-dispossessed homeless people from viable solutions. Land owners could apply to a subsidy for upgrading their house or for a new one on-site, but the new collective housing or housing purchasing programmes are, in the current situation in most cases, inapplicable to the dispossessed group, leaving them no choice but to increase informality. Therefore, the elimination of differences has the potential to result in a process of exclusion, moreover relying on the existing operative tools for housing overburdened existent institutional capacities. As an illustration, local councils could not cope with increasing demands and resources, and in many cases, got caught in bureaucratic processes. The scale of the disaster required more adaptable and multiple alternatives in which the affected population could effectively customize their solutions.

Nevertheless, one of the measures proposed seeks to achieve reconstruction on a broader scale than housing, pursuing a more integral view of recovery than simply disconnected micro-localized housing projects. For almost completely destroyed areas, such as coastal zones devastated by the tsunami, the idea was to generate master plan proposals aiming for a more integral urban design process. This was imagined through the definition of 10 *Sustainable Reconstruction Plans* (PRES) in major cities and 102 *Urban Regeneration Plans* (PRU) in intermediate cities (PNUD, 2010). The bidding process for the first 20 pilot PRU's plans' consultants is managed by a joint-venture between MINVU and the United Nations Development Programme (UNDP). The funding for those consultancies is provided by the UNDP in Chile (i.e. PNUD) as after 2005 earthquake the UN promoted an interagency programme of cooperation for coping disasters in Chile (PNUD, 2010). Due to the pilot-project nature of this plans, it remains unclear as to how they will be developed and implemented in order to create sensible responses to the needs of the displaced population and transform disaster into an opportunity to 'build back better' (Lyons et al., 2010). Beyond the rhetorical debate around the notion of build back better, which is rapidly becoming a new post-disaster slogan, many questions arise in relation to the nature of the master planning logic of reading the situation and imagining the future of the affected Chilean cities. What are the ways to produce an effective implementation of these plans? How should they deal with land speculation once the plans are completed? How long might these plans be sustained? If the plans are developed without a coherent grasp of the real and urgent needs or not aligned with current urban planning procedures, the result might be a mere abstract representation of reality.

Specifically in Constitución, the professional consultants developing the PRES, proposed a set of integral urban strategies concerning disaster risk reduction, housing, public space and infrastructure. Alejandro Gutierrez, an *Arup* expert states 'the plan fundamentally aims to build a better city than the previous Constitución' (2010), and drawing on that the plan must propose solutions to previous issues such as transport congestion in the city centre, continuous river flooding, risk of tsunamis, affordable energy access, etc. (Gutierrez, 2010). The design mitigation measures to future floods proposed (e.g. tsunamis, marsh floods, river floods, etc.), respond to experts views on the need to mitigate possible impacts caused by new tsunamis, earthquakes or river flooding. The urban design proposal comprises a flood park in the coastal area and a new system to canalize the marsh bed. The proposal also considers the implementation of safety measures for future events such as evacuation streets with photovoltaic illumination, markers, sirens and the demarcation of secure meeting points (PRES, 2010c). Ninety-four per cent of the citizens approved at the public consultation the flood park proposal (PRES, 2010e). However, residents of the coastal areas that would be displaced due to the park creation might be represented in the remaining percentage (Figure 4). In order to find alternatives solutions for those directly affected, special working meetings were held in the open house (PRES, 2010f). This urban proposal goes beyond the mere production of houses intending to prepare the city to future unforeseen events. However, the effectiveness of these plans will be limited to a graphical representation if the experts do not encourage an active involvement of the citizens in the implementation of these measures. The post-disaster housing demand of about 2,000 units over the existent pre-disaster housing deficit (around 3,000 units) (PRES, 2010c), pose to the master plan additional complexities, which were tackled specifically by suggesting an increased density in the central area (i.e. from 30 dwellings per hectare to 200 dwellings per hectare). However, land availability in the centre might be detrimental to the feasibility of these plans. The housing response has gone slower than expected and currently under the master plan framework only two projects of 300 housing solutions each designed by ELEMENTAL – located in the proposed areas of urban extension rather than in the centre – are under development (i.e. one being built and the other in the process of approval). Besides,

FIGURE 4 View of the coastal area affected by the Tsunami; December 2010.

independently from the master plan proposals, four collective housing projects– coordinated by other organisms (EGIS) – are under planning and regulation process. Together, these projects will account for around 1,000 solutions, and following the same trend only 70 of these solutions will be situated in the central area (MINVU, 2011a). How does the densification of the central area, drawn by the master plan proposal, fit into this trend? How effectively are these proposals integrated in the re-construction of the city? Considering these facts seems apparent the need to broaden the alternatives for addressing the housing needs of the affected as to achieve an equitable process.

The adoption of master planning solutions, though praised for incorporation in the pre-disaster mode of production of spaces and cites in terms of policies, normative and operational frameworks, not only requires a specific set of knowledge but also an important set of attention, sensibility and thus interpretation of the realities and the potential effect of such a plan on the future. Moreover such a transformative strategy, if following Sanderson (2010) when stating that architects are the last people needed in disaster reconstruction, denies their potential for being an 'interpretive agent' (Till, 2009, p. 164), leaving no 'task for [a] translator' (Wigley, 1988, p. 660). Hernández argues that 'as language change[s], new translations will be needed' (2010, p. 27). As with language, the persistent adaptation of urban space and particularly the rapid changes that occur due to the destruction of devastating unforeseen events, do demand translators and need translation. The disciplines of architecture and urban design cannot be set apart. Rather they are needed to raise their 'significant power to reconstruct social networks, raise solidarity, empower communities and encourage partnerships' (Boano & Hunter, 2010). Master planning representation and imagination carry the risks of not being translated if, as often happens, there is only a 'literary application' of the formal mode of production of cites and plans a 'material representation of an abstract idea' (Wigley, 1988, p. 660).

9. Conclusion: '(De)colonizing' reconstruction

What emerges from the Chilean case is that the Government, while attempting to guide and direct the massive task of reconstruction, refuses to acknowledge the diverse needs of different cultural groups, leaving many excluded from the recovery process. Somehow, the government vision of problem solving in the 2010 Chilean earthquake case, visible in their proposal '7 solutions for 7 types of problems' (MINVU, 2010b), denies the possibility of agency promotion. Quoting Till, 'it is difficult to reconcile the notion of transformative agency with that of problem-solving. Problems look determinedly backward, while agency looks hopefully forward' (Till, 2009, p. 167).

The adoption of an overly simplistic reading of reality, reducing existing needs into seven types with seven solutions, fails to recognize the complexity of a post-disaster situation and denies the relevance of cultural differences, bringing into sight the domination of colonial discourse. Rather, reconstruction solutions must be open, transformative and adaptable to respond to 'cultural differences' (Bhabha, 2004, 2009), and dynamic processes over time. Due to the complexity of a disaster scenario, the 'urban' gains importance, becoming a space for negotiation in considering the multiple interests involved. Architecture and architects should understand the connections between the societal processes of living in a city and their spatial manifestations. Writing off the spatial element as something separate and unrelated severely limits the transformative potential of reconstruction and rehabilitation in the post-disaster context. Architecture as a discipline and practice will continue to be at risk as long as it fails to perceive its own critical reflections and its object obsession, while recognizing its dependency on outside forces and influences.

Translation transforms the original being into a re-creation of it (Wigley, 1988; Bhabha, 2004; Hernández, 2010). However, when translation is reduced to a 'literary application' of predetermined solutions, complex abstractions and imaginations, it ceases to be a genuine translation (Wigley, 1988; Hernández, 2010), but rather exists as a 'material representation of an abstract idea' (Wigley, 1988, p. 660). In doing so, the translator – or in this case, the architect – has been focused on the 'object of intention' rather than in the 'mode of intention' (Hernández, 2010, p. 29); that is in the physical representation instead that in the meaning and objective of the translation. Therefore, the architect becomes *lost in translation*. Referring back to the table of translation and representation, the critically posed questions aim to navigate the path of the architect in such a complex professional terrain.

Indeed, the conceptual apparatus of this study and its adaptation to the case of post-disaster Chile seek to begin a post-colonial reading of reconstruction, which has been strangely absent in the literature. Aiming to repopulate and reinvigorate the call for a radical thinking on the powers of architecture and urbanism, by denoting the differences between colonized and colonizer, ratifies the superiority of the colonizer, detaching the colonized from its own history, legacy and identities, The imposition of dominance and power within colonialism neglects spatial agency. On the contrary, inspired by the recent work done by the Bethlehem practice of Hilal et al. (2010) (de)colonizing reconstruction can potentially engender the capacity to translate and represent a different praxis of post-disaster reconstruction, liberating it.

As Till argues such urgent move is 'from a [...] impulsive imagination to [...] collaborative ethical imagination' (2009, p. 151). If this critical separation can be achieved then space will be opened for more culturally sensitive approaches to the production of post-disaster space, where architects and designers are needed to translate and not become lost in translations. The criteria and indicators' previously identified aim to provide the space of critical separation needed to engage in such collective imagination. In not giving up on Architecture's capacity for resistant affirmations, we can reduce the risk of its operations in post-disaster practice being limited to mere autonomous, service-oriented provision.

Acknowledgement

We would like to thank Andrew Wade for reading the various versions of this paper and providing valuable feedbacks, editorial suggestions and encouragement. We would also like to thank the reviewers whose comments have improved this contribution considerably.

References

Aliaga, F., 2010. *Plan para reconstruir Constitución costará más de US$ 1 millón*. EMOL. 20 de abril 2010. www.estrategia.cl/detalle_noticia.php?cod=31485&sec=12 (accessed 23 July 2010).

Appadurai, A., 2003. Illusion of permanence interview with Arjun Appadurai. *Perspecta*, 34. 44–52.

Benjamin, W., 1968. *The Task of the Translator*. In Hannah Arendt, *Iluminations*. Harry Zohon (trans.). Schocken Books, New York.

Bhabha, H., 2009 (1988 original). Cultural diversity and cultural differences. *The Post-Colonial Studies Reader* (1st edn, 1995). B. Ashcroft, G. Griffiths and H. Tiffin (eds). Routledge, Oxford and New York.

Bhabha, H. K., 2004. *The Location of Culture*. Routledge, New York.

Blunt, A. and McEwan, C. (ed.), 2002. *Postcolonial Geographies*. Continuum Studies in Geography, London and New York.

Boano, C., 2009. Housing anxiety, paradoxical spaces and multiple geographies of post tsunami housing intervention in Sri Lanka. *Disasters*, 34(3). 762–785.

Boano, C. and Hunter, W., 2010. Risks in post disaster housing: architecture and the production of space. *ABACUS International Journal on Architecture, Conservation and Urban Studies*, 5(2). 23–31.

Bresciani Lecannelier, L. E., 2010. Chile 27F 2010: La catástrofe de la falta de planificación. *EURE (Santiago)*, 36(108). 151–153. www.scielo.cl/pdf/eure/v36n108/art07.pdf.

Cancino, A., 2010. *Blog entry PRES Constitución*. http://presconstitucion.cl/noticias/foro-de-arauco-en-el-pres/ (accessed 26 July 2010).

Dovey, K., 1999. *Framing Places: Mediating Power in the Built Form*. Routledge, London.

Dovey, K., 2010. *Becoming Places. Urbanism/Architecture/Identity/Power*. Routledge, London and New York.

El Mercurio, 2010. *Video at Constitución tras la catástrofe. Bitacora de una Reconstrucción*. Capítulo 8, Un año sin soluciones definitivas. www.emol.com/especiales/2010/nacional/constitucion/index.asp (accessed 27 February 2011).

FNR (Fondo nacional de la Reconstrucción), 2010. *Portafolio de Proyectos por Región*. http://donaciones.hacienda.cl/portafolio-de-proyectos/por-region.html.

Gutierrez, A., 2010. *Interview to Alejandro Gutierrez, ARUP consultant*. www.presconstitucion.cl/noticias/page/28/ (accessed 30 March 2011).

Hilal, S., Petti, A. and Weizman, E., 2010. *Decolonising Architecture Art Residence*. www.decolonizing.ps/site/?page_id=2 (accessed 8 December 2010).

Hernández, F., 2010. *Bhabha for Architects*. Routledge, Oxford.

Hormazabal, A., 2010. *Community leader La Posa, Constitución. Interview in video Blog PRES*. http://presconstitucion.cl/video-blog/?tubepress_page=6.

IFRC, 2010. *World Disaster Report 2010 Focus on Urban Risk*. International Federation of Red Cross and Red Crescent Societies, Lyons.

INE, 2010. *Proyecciones de población*. www.ine.cl/canales/menu/indice_tematico.php.

Jacobs, J. M., 1996. *Edge of Empire. Postcolonialism and the City*. Routledge, London and New York.

Jha, A. K., Duyne Barenstein, J., Phelps, P. M., Pittet, D. and Sena, S., 2010. *Safer Homes, Stronger Communities. A Handbook for Reconstructing after Natural Disasters*. Global Facility for Disaster Reduction and Recovery. World Bank, Washington.

Lefebvre, H., 1991. *The Production of Space* (1974, 1st edn). Translated from the French by D. Nicholson-Smith. Oxford, Wiley-Blackwell.

Lefebvre, H., 1997. The production of space (extracts). *Rethinking Architecture. A Reader in Cultural Theory*, N. Leach (ed.) (1997). Routledge, London.

Lizarralde, G., Johnson, C. and Davidson, C., 2010. *Rebuilding after Disasters. From Emergency to Sustainability*. Spon Press, Oxford and New York.

Lyons, M., Schilderman, T. and Boano, C. (eds), 2010. *Building Back Better. Delivering People-Cetred Housing Reconstruction at Scale*. Practical Action Publishing, London.

McFarlane, C., 2004. Geographical imaginations and spaces of political engagement: examples from the Indian Alliance. *Antipode*, 36(5). 809–916

MINVU, 2010a. *Chile Unido Reconstruye Mejor*. www.gobiernodechile.cl/informa/2010/03/31/programa-chile-unido-reconstruye-mejor.htm (accessed 6 July 2010).

MINVU, 2010b. *Tríptico plan de reconstrucción nacional*. www.minvu.cl/opensite_20100506111833.aspx (accessed 6 July 2010).

MINVU, 2010c. *Resolución Exenta 1878 del 26 Marzo 2010*. www.minvu.cl/aopensite_20100910140027.aspx (accessed 25 March 2011).

MINVU, 2011a. Reconstrucción Nacional MINVU. Presentación Región del Maule Proyectos de Vivienda. January 2011.

MINVU, 2011b. Plan de Reconstrucción. www.minvu.cl/opensite_20101001180448.aspx.

ONEMI, 2010. *Noticias*. www.onemi.cl/index.html (accessed 15 June 2010).

PLATAFORMA URBANA, 2010. www.plataformaurbana.cl/archive/2010/06/17/presconstitucion-vecinos-de-constitucion-aprueban-parque-fluvial-maule-y-priorizan-otros-cinco-proyectos-para-el-plan-maestro/.

PNUD, 2010. *Planes de Regeneración Urbana para la VII, VIII y IX Región*. www.pnud.cl/servicios/1.asp.

PRES, 2010a. *Web page and Forums*. http://presconstitucion.cl/ (accessed July 2010).

PRES, 2010b. *Minuta primera consulta ciudadana abierta*. http://presconstitucion.cl/wp-content/uploads/2010/04/Minuta-primera-consulta-ciudadana-abierta.pdf (accessed 18 July 2010).

PRES, 2010c. *Presentación Final Plan Reconstrucción Sustentable*. http://presconstitucion.cl/page/2/.

PRES, 2010d. *Pres Constitución avanza en su implementación*. 30 de Agosto. http://presconstitucion.cl/noticias/page/5/.

PRES, 2010e. *PRES Constitución presenta a Ministra de Vivienda avances del plan de reconstrucción*. 03 de Junio. http://presconstitucion.cl/noticias/page/14/.

PRES, 2010f. *Mesa de trabajo en el PRES*. 31 de Mayo. www.presconstitucion.cl/noticias/page/18/.

PRES, 2011. *Ver índice de lugares*. www.presconstitución.cl (accessed 15 March 2011).

Sanderson, D., 2010. *Architects are Often the Last People Needed in Disaster Reconstruction. The Guardian*, 3 March 2010. www.guardian.co.uk/commentisfree/2010/mar/03/architects-disaster-reconstruction-haiti-chile (accessed 5 March 2010).

Sharp, J., 2009. *Geographies of Postcolonialism*. Sage Publications, London.

Sierra, L., 2006. Cap. 11. Urbanismo por Decreto: Centralismo y Confusión Institucional en la Ciudad Chilena. *Santiago: Dónde Estamos y Hacia Dónde Vamos*, A. Galetovic (ed.). Centro de Estudios Públicos, Santiago, 2006.

Spivak, G., 1999. Can a subaltern speak? *The Post-Colonial Studies Reader*, B. Ashcroft, G. Griffiths and H. Tiffin (eds) (2009). Routledge, Oxford and New York.

Till, J., 2009. *Architecture Depends*. MIT Press, Cambridge.

Tillería, H., 2011. *Interview to CNN Chile*. www.cnnchile.com/nacional/2011/02/22/postterremoto-nos-faltan-mas-de-5000-viviendas-en-constitucion/.

Valencia, M., 2010. *Gobierno ha entregado 10 mil viviendas a los damnificados del terremoto. In La Tercera (2011)*. http://diario.latercera.com/2011/02/27/01/contenido/pais/31-60802-9-gobierno-ha-entregado-10-mil-viviendas-a-los-damnificados-del-terremoto.shtml (accessed 27 February 2011).

Wigley, M., 1988. The Translation of Architecture, the Production of Babel. *Architecture Theory since 1968*, M. Hays (ed.) (2002). New York, Columbia University and MIT. 658–675.

Zetter, R., Boano, C. and Lizarralde, G., 2010. Space and place after natural disasters and forced displacement. *Rebuilding after Disasters. From Emergency to Sustainability*, G. Lizarralde, C. Johnson and C. Davidson (eds) (2010). Spon Press, Oxford and New York.

Is there a human right to shelter after disaster?

RICHARD CARVER

Centre for Development and Emergency Practice (CENDEP), School of the Built Environment, Oxford Brookes University, Gipsy Lane, Oxford, OX3 0PB, UK

There is understandable scepticism about the introduction of human rights language into disaster response. In contrast to the use of rights-based approaches to development programming, there has been relatively little attempt to use human rights as a normative or policy framework for the provision of shelter after disaster. Human rights standards contain a highly developed definition of the right to adequate housing, yet in its fullest iteration this is not of particular relevance for immediate post-disaster action. However, the seven elements in the human rights definition of adequate housing become highly pertinent in the transition from disaster response to longer-term reconstruction and development. The most important contribution of human rights to shelter after disaster, however, may lie elsewhere. Broader human rights norms stress the importance of equality and non-discrimination in the provision of shelter and highlight the danger of making trade-offs between shelter rights and other rights, such as freedom of movement or access to livelihoods. 'Human rights-based approaches to programming' have a tendency to overstate the contribution of human rights in these contexts, often failing to grasp that the prism of rights is not the only approach to all human problems, or always the most useful. Yet, a rights-based approach that is firmly anchored in internationally agreed standards can be an invaluable way of returning the issue of state obligations in relation to adequate housing to the centre of the debate.

1. Introduction

There is considerable and understandable scepticism about the introduction of human rights language into the arena of disaster response. 'Human rights' has become the dominant discourse in international relations in the post-Cold War world and it is reasonable to ask whether the use of rights talk in an area such as shelter after disaster is merely the application of a fashionable and dominant vocabulary. The onus, then, is on human rights advocates to first explain what are the relevant norms that apply to shelter after disaster and then, even more importantly, what practical difference the adoption of a human rights approach might make.

It is hard to see how the bricks and mortar – or timber and canvas – of shelter after disaster would be different if a human rights-based approach were adopted, although there are rather well-developed standards to define the right to adequate housing in a broader sense. What human rights can do, however, is provide an understanding of the broader entitlements of those who are displaced in disasters. The right to adequate housing might not even be the most relevant human rights norm to take into account in post-disaster shelter programming. Rather, provision of shelter after disaster should be informed by the understanding that the displaced and homeless continue to be entitled to their whole range of human rights.

A human rights approach would require, for example, that shelter be provided without discrimination on the basis of need. Nor can provision of shelter be made conditional, explicitly or implicitly, upon the surrender of other rights, such as freedom of movement. Shelter would also have to be provided in a manner that recognizes the importance of human rights such as the right to a livelihood in displaced people's reconstruction of their lives. Human rights standards set out seven criteria that go to define the right to adequate housing, which are discussed in this article. What is noticeable in these criteria is the strong emphasis on links between the right to housing and other rights, such as water and sanitation, livelihood, health and education. Human rights do not propose a maximalist programme in which all elements of the right to adequate housing must be present in shelter after disaster. Such an approach would be impractical and condemn human rights advocates to irrelevance. Yet, the definition of adequate housing contained in human right standards can usefully inform the provision of post-disaster shelter.

It is argued in this article that a human rights approach becomes particularly valuable in the transition from shelter as an immediate emergency response into longer-term reconstruction or development. In the first phase, the leeway for external agencies is limited and governments may find themselves close to powerless. Moving beyond shelter after disaster, however, entails clear policy choices by all actors, with national authorities much more likely to make decisions that will affect future enjoyment of human rights. It is at this point that the seven criteria that comprise the human right to adequate housing can be realistically integrated into plans for the construction of intermediate or long-term housing for the displaced. Above all, using the language of citizen entitlement may help to focus the attention of policymakers on the nature of their obligations.

To answer the question posed in the title of this article: of course, in a strictly legal sense there is a human right to shelter after disaster. There is a human right to adequate housing, enshrined in international law and broadly accepted. If someone's home is destroyed, this does not mean that they lose the right to housing – indeed, if human rights have any meaning, it makes the fulfilment of that right rather more urgent. The more relevant questions therefore are the following: what does the right to housing in the aftermath of a disaster consist of? And in what sense, if any, is this a useful way of looking at the provision of shelter in emergencies?

This article approaches these questions in four steps. First, it looks at the content of the right to adequate housing and its applicability in a post-disaster situation. Second, it discusses the question through the prism of the rights-bearers, the internally displaced who belatedly have acquired a distinct status within international human rights norms. Although the internally displaced enjoy, in principle, the same rights as everyone else, they may face particular obstacles to the enjoyment of these rights, as do those returning to their homes in the aftermath of disaster. Third, the article considers how far a 'human rights-based approach to programming' may be an effective tool for planning and implementing the provision of shelter after disaster. It looks in particular at the strengths and weaknesses of the approach adopted by several UN agencies. Fourth, there is a brief look at two recent examples of disasters that caused massive homelessness – the 2004 tsunami in Sri Lanka and the 2010 earthquake in Haiti – observing how human rights monitoring played an important role in the former instance, but not in the latter. Finally, a concluding discussion draws these threads together, offering suggestions both on the contributions that human rights can make to shelter after disaster and on their limits.

2. The right to adequate housing

The right to adequate housing in Article 25 of the Universal Declaration of Human Rights (UDHR) is part of a broader right to an adequate standard of living that also encompasses food, clothing, medical care and social services. A similar package is contained in Article 11 of the International Covenant on Economic, Social and Cultural Rights (ICESCR), which gives binding force to the right for those states that are party to it.

Other human rights treaties addressing particular social groups – including women, children, migrants and people with disabilities – also make explicit mention of this right, which hence also has binding force on states that are party.

In General Comment 4, the Committee on Economic Social and Cultural Rights, the treaty body charged with monitoring and interpreting the ICESCR, set out a non-exhaustive list of seven elements that comprise *adequacy* in relation to housing: legal security of tenure, availability of services, affordability, habitability, accessibility, location and cultural adequacy[1]:

- Forced eviction is regarded as the most serious *violation* of the right to adequate housing. Legal security of tenure, whatever the status of the resident, is a necessary protection against forced eviction and harassment.
- In order to meet the criterion of adequacy, services, materials, facilities and infrastructure should be available. Adequate housing is thus not just a roof over one's head, but also the availability of safe water, sanitation, energy for heating and cooking, refuse disposal and emergency services.
- The standard of affordability requires that housing, including necessary services, be commensurate with income levels, and that states ensure that housing finance be available, including in the form of subsidies, to those who could not otherwise afford housing.
- Habitability requires that housing have sufficient space and be structurally sound enough to provide shelter and protection against the elements and disease vectors.
- Accessibility means that adequate housing must be available to all those entitled to it. The needs of disadvantaged groups are particularly relevant here, including children, the elderly, people with disabilities and female-headed households. In its discussion of the accessibility standard, the Committee made particular reference to victims of 'natural disasters' and those living in 'disaster-prone areas' as groups entitled to priority consideration in provision of housing.
- The location standard addresses the need for housing to be available in places where there are opportunities for employment or other economic activities, as well as health care, education, transport and other social facilities.
- Finally, the standard of cultural adequacy refers to vernacular building styles and materials used, in order that housing be acceptable to the population. There may also be considerations of privacy and space that are particular to any given culture or society.

Read together, these seven criteria constitute a comprehensive check-list for the adequacy of housing. How far they are of practical relevance in a post-disaster context may be a different matter.

The ICESCR, in Article 2, makes a general assumption that economic and social rights are subject to 'progressive realization.' This is usually understood to mean that the economic constraints on instant fulfilment of these rights mean that they can only be realized gradually. In a disaster situation, of course, the constraints may be not merely economic but also practical and logistical. The sheer scale of the sudden need to provide shelter (and many other services) to large numbers of people means that there can be no realistic obligation on governments to provide 'adequate housing' in its fully elaborated sense.

The precise obligations of governments within the context of progressive realization have been the subject of considerable debate and elaboration. Two paradigms are of particular relevance. The first is the assumption that governments' obligations in relation to any particular right fall under three distinct headings: to respect, protect and fulfil. The obligation to respect is essentially negative – the state must not interfere with the individual's enjoyment of a right. The obligation to protect relates to the acts of third parties – hence, the state must ensure that an individual's rights are not infringed by other actors. The obligation to fulfil entails positive steps to meet the full enjoyment of a right. It is immediately apparent in relation to the seven standards for adequate housing that, for example, security of tenure falls within the spheres of respect and protection, requiring states not to carry out forced evictions and to protect individuals from forced evictions by others, such as landlords or developers. Most of the other elements fall largely within the obligation to fulfil, to the extent that they entail the development of infrastructure, the construction of housing and so on.

The second paradigm is that each right has a 'minimum core content'. This is the irreducible minimum that must be met from the outset, even within the constraints of progressive realization. It says, in essence, that there are some aspects of a right that impose a binding obligation on states in all circumstances, while others will only be fulfilled over time. The concept of minimum core content is more controversial than the three-legged respect–protect–fulfil paradigm. Critics fear that the minimum core content becomes the essence of the right, while other aspects are seen merely as optional extras. The notion of progressive realization itself is clearly open to the same problem, with the risk that substantive progress will be postponed indefinitely. The Indian Supreme Court had harsh words for successive governments that had failed for decades to meet social and economic goals established in the Constitution, suggesting that at a certain point realization could no longer be progressive and became an immediate duty: 'Does not the passage of 44 years... convert the obligation created by the article into an enforceable right?' (Unnikrishnan, 1993).

3. The housing rights of internally displaced persons

International human rights law approaches the guarantee of rights in two distinct ways. One, already described, is to elucidate the content of rights. The other is to look at rights-holders. The key United Nations human rights treaties on civil and political, and economic, social and cultural rights are supplemented and elaborated by a series of further treaties that set out rights for specific vulnerable groups, including women, children, migrant workers and people with disabilities. Other standards – such as declarations and codes of practice – deal with the rights of other groups, including prisoners, indigenous peoples and the elderly.

These human rights instruments do not create new rights, but rather address obstacles that are specific to each group in the enjoyment of universal human rights. In post-disaster emergencies the vulnerability of those affected consists mainly in the fact that they have been displaced (although it may rest also in their social or ethnic background, age, sex and so on). Most people made homeless by disasters can be regarded as internally displaced persons (although a small minority are refugees). There is no human rights treaty explicitly providing for the rights of internally displaced persons, although this gap in international protection has been addressed in recent years. Before 1992, there existed no agreed international definition of internally displaced persons (IDPs), who were defined essentially negatively – they were people who had been displaced but were not refugees, because they had not crossed an international border (Phuong, 2004). In 1998, the UN Commission on Human Rights adopted the Guiding Principles on Internal Displacement, which set out the rights of IDPs. While this document has no binding force – and has to contend with the hostility of governments that tend to resent what is seen as interference in their handling of an internal issue – the Guiding Principles are formulated primarily as a restatement of existing international law, drawing on human rights and international humanitarian standards. (The Principles exist in an annotated edition that exhibits an admirable and unusually scrupulous attention to the exact sources of the rights enunciated (Kälin, 2000).) Gould (2009) notes that the Principles were formulated to deal with those displaced by conflict and that those affected by disaster have been the 'stepchild of conflict law', addressed in a somewhat perfunctory manner that does not deal with the specificities of their situation.

In relation to housing, the Principles identify 'basic shelter and housing' as one of the minimum provisions that governments shall ensure 'regardless of the circumstances, and without discrimination'. They explicitly add that 'Special efforts should be made to ensure the full participation of women in the planning and distribution of these basic supplies'.

The Principles contain a couple of other points of potential relevance to the provision of shelter after disaster. One is a prohibition on arbitrary deprivation of property. The other is a general presumption against the housing of IDPs in camps where their movement will be restricted.

For those engaged in humanitarian assistance after disasters, however, potentially the most important section of the Principles is that addressing the conduct and obligations of humanitarian organizations. In particular, Principle 27 states in part:

> ... International humanitarian organizations and other appropriate actors when providing assistance should give due regard to the protection needs and human rights of internally displaced persons and take appropriate measures in this regard. In so doing, these organizations and actors should respect relevant international standards and codes of conduct.

The importance of this is that it binds together two roles for international organizations that in their genesis and history have been generally distinct: humanitarian assistance and protection. The implications of this are discussed below.

It was within the framework of protection of IDPs that the Inter-Agency Standing Committee produced a set of *Operational Guidelines on Human Rights Protection in Situations of Natural Disaster*. These were elaborated essentially to supplement the Sphere Guidelines, which were regarded as being insufficiently stringent and rights based.[2] The rights contained within these Guidelines are, in essence, the same as those in the Guiding Principles on Internal Displacement. The Guidelines spell out some of the operational implications of the Principles, focus exclusively on 'natural' disasters (those not resulting from conflict) and, perhaps most importantly, distinguish the varying human rights issues that are likely to emerge in the different phases of post-disaster emergencies. In immediate post-disaster response, 'The right to shelter should be understood as the right to live somewhere in security, peace, and dignity'. Shelter in this phase, in common with other basic human entitlements, shall be 'considered adequate if they ensure survival to all in need of them'. That is, the entitlement is minimal and basic, but it is enjoyed on a non-discriminatory basis, as need dictates, and it is confined to the emergency phase. Four general standards are set out for the provision of basic requirements in the immediate emergency phase: availability, accessibility, acceptability and adaptability. (Sharp-eyed readers will have noticed that those who draft UN human rights standards favour abstract nouns beginning with 'a' and ending with 'y'.) Availability means that services are available to the affected population in sufficient quantity and quality. Accessibility refers both to physical proximity and provision without discrimination to all who need the relevant services. Acceptability refers to the cultural appropriateness of the services provided. Adaptability means that the goods and services are appropriate to be used in the various phases of the emergency response, including relief, reconstruction and return.

The Guidelines also make a priority of a speedy transition from the shelter phase to the provision of 'temporary or permanent housing' that meets the international standards of adequacy – defined according to the seven criteria outlined above. The Guidelines also stress the importance of involving affected people in planning new housing developments, through consultation and participation. Particular mention is made of women, indigenous people and people with disabilities.

The Guidelines state that owners of destroyed houses should, as far possible, be able to decide themselves how they are rebuilt. There is also recognition that the issue of proving title to property may be particularly difficult in a disaster, when documents are often lost or destroyed. The issue of restitution of property may be important. The Guidelines require that there be access to impartial court proceedings in the event that disputes cannot be settled otherwise. They recognize that there may be particular problems for groups such as women to establish title to property or for sections of society, such as indigenous peoples, that may not recognize individual title to land. The Guidelines require steps to be taken to protect those affected, particularly poor and vulnerable groups, from deprivation of property by landlords, speculators and local authorities.

Prohibitions on returning should be based on law and should only apply where necessary for reasons of health, safety, or disaster prevention, or for purposes of reconstruction and development.

Finally, the Guidelines recognize that there may be circumstances in which evictions are required to guarantee the safety of people. A series of safeguards are recommended for the conduct of evictions:

> (a) an opportunity for genuine consultation with those affected; (b) adequate and reasonable notice prior to the scheduled date of eviction; (c) the timely provision of information on the eviction and future use of the land; (d) the presence of government officials during an eviction; (e) the proper identification of all persons carrying out the eviction; (f) the prohibition of evictions during bad weather or at night; (g) provision of legal remedies; and (h) provision of legal aid, where needed, to seek redress from the courts.

A problem of particular relevance to post-disaster housing is how returning displaced people gain access to property that may be occupied by others or have been purportedly reallocated legally. As an international legal problem this is normally thought of in the context of conflict, as in former Yugoslavia, annexation, as in Cyprus, or the sequel to the downfall of authoritarian regimes, as in a number of post-Communist countries. But the reallocation of land from previous residents to developers has become a common problem in the aftermath of disaster (Housing and Land Rights Network (2006); Advisory Group on Forced Evictions (2010) and Klein (2007)). In 2001, the UN Sub-Commission on the Promotion and Protection of Human Rights appointed a Special Rapporteur on Housing and Property Restitution for Refugees and Internally Displaced Persons. In 2005, he produced a set of principles – known by the rapporteur's name as the 'Pinheiro Principles'. As with the Guiding Principles on Internal Displacement, these Principles consolidate and codify previously existing international law on this issue, as well as using these existing principles to create more concrete requirements for the various actors involved, primarily states (Pinheiro, 2005).

4. Human rights-based approach to programming

What has been described so far is some overlapping sets of normative standards, of varying levels of authority, that may be relevant to the provision of shelter after disaster. The *Operational Guidelines on Human Rights Protection in Situations of Natural Disaster* may be of the greatest practical application for agencies on the ground, precisely because they are intended to be operational. They are also available integrated into a field manual that gives further guidance on the operational implications of the human rights standards contained therein.

In recent years the attempt to integrate human rights into development has seen the formulation of 'human rights-based approaches to programming' (HRBAP) (Cornwall and Nyamu-Musembi, 2004; Darcy, 2004; Gready and Ensor, 2005). (While this is usually described as a human rights-based approach to *development*, there is no reason as to why this would not apply equally in an emergency context and, in any case, an effective emergency response should also take account of the transition to the developmental phase.) The human rights-based approach is available in multiple varieties, the most widely used and influential being those developed by the UN agencies UNDP, UNICEF and UNFPA.[3] The UN approach has its most authoritative statement in the 2003 Statement of Common Understanding on HRBA among UN Agencies.

The UN HRBAP is based on the assumption that the *process* of development and emergency response is as important as the *outcome* and that human rights should guide both of these. 'A human rights approach requires equal attention to outcome and process' (Jonsson, 2005). Outcomes are determined by human rights standards. This, in part, implies the hard standards of positive international law. According to one of the architects of this approach, however, positive law is:

> ... too dogmatic and not very useful for an HRBAP. It would also miss the point that ... new rights will be constructed, gradually codified in conventions and accepted by ratification (Jonsson, 2005).

Good process is conducted in accordance with human rights principles. According to the UN Common Understanding these are:

- universality and inalienability
- indivisibility
- interdependence and interrelatedness
- equality and non-discrimination
- participation and inclusion
- accountability and rule of law.

The UN's HRBAP relies in addition, and very importantly, on an understanding of a web or pattern of claim–duty relationships. Whereas the traditional approach to human rights, grounded in international law, sees obligations attaching to the state, the HRBAP identifies obligations on the part of many actors. Being a rights holder or a duty bearer is not a permanent attribute but a functional role. Hence, any individual might simultaneously hold rights and bear duties (Jonsson, 2005). The reason why these relationships are central is that the purpose of all programming within the rights-based approach is to strengthen the capacity of rights holders to claim their rights and of duty bearers to meet their obligations.

In practice, a number of steps are entailed in programming using this method, including identifying underlying causes for human rights violations, analysing the claim–duty relationships, determining where and why capacity falls short and planning steps to fill this capacity gap.

The HRBAP (whether in a development or emergency context) brings one important and undeniable advantage, but at the same time raises several problems. On the positive side, the HRBAP emphasizes that *all* human rights apply at all times. In the present example, it is important that in providing shelter after disaster all other human rights are respected, not only rights of non-discrimination that may relate directly to shelter, but other civil and political rights that may be affected by, for example, the decision to house displaced people in camps. This point may seem self-evident, but it has certainly not been reflected in the practice of those international agencies historically most involved in provision of shelter for the displaced, notably the United Nations High Commissioner for Refugees (Harrell-Bond, 1986).

On the other hand, however, there is an intellectual incoherence to the HRBAP that leaves it open to criticism.

The definition of human rights does not correspond to any broadly accepted international definition: To the extent that the human rights standards invoked are those contained in the international treaties and other standards, they are unexceptionable. When these standards are something that fall short of internationally agreed human rights, they move into the area of wishful thinking and raise the question of why it is necessary to frame these social benefits in rights terms. The Common Understanding states that:

> All programmes of development co-operation, policies and technical assistance should further the realisation of human rights as laid down in the

> Universal Human Rights and other international human rights instruments (Common Understanding, 2003).

And it clarifies that this does not mean programme activities that 'only incidentally' contribute to the realization of human rights, but rather require that 'the aim of all activities is to contribute directly to the realization of one or several human rights'. This clearly makes no sense at all within the usual definition of these terms. To take an obvious and relevant example, a programme promoting disaster risk reduction (DRR) is not aimed at realizing human rights (except in the most indirect way). Either we must conclude that the UN will not undertake DRR, or that there is a 'right to DRR'.[4] This is the sort of sloppy thinking that makes professionals in the disaster and emergency field sceptical of human rights.

The human rights principles that guide 'process' are not derived from internationally agreed definitions either: In the list of six 'human rights principles' cited above, it is notable that only three of them are unambiguously contained in international human rights documents: universality, interdependence, and equality and non-discrimination. Indivisibility is a human rights principle, but it does not mean what the Common Understanding says that it means, which is that all human rights 'have equal status as rights, and cannot be ranked a priori in hierarchical order'. Lest it be thought that this is simply poor drafting, and that the intention was to say that no *group* of rights was inferior – civil and political or economic, social and cultural – the Frequently Asked Questions on the HRBAP produced by the UN High Commissioner for Human Rights states clearly: 'The principle of their indivisibility recognizes that no human right is inherently inferior to any other'. Evidently this is just nonsense. Saying that, for example, the right to social security is equivalent to the right not to be held in slavery is likely to bring human rights into disrepute. Of course, human rights law in reality says no such thing. Some human rights – such as the right not to be enslaved – are overriding or peremptory norms of international law, others may be suspended in times of emergency, and many will have to be balanced against each other in their actual implementation. Many practitioners are deterred from referring to human rights in an emergency context because of precisely such a belief that they are inflexible and never subject to compromise. It is unhelpful for the UN itself to reinforce that impression.

Participation and inclusion are also problematic as human rights principles. The UDHR and the core human rights covenants do establish a right to participation, but in a very specific context: the right to participate in the government of one's country through periodic elections, to be eligible to work in the public service, and to participate in 'public affairs' either directly or through elected representatives. None of these – which are primarily to do with a right to representation – is related to 'participation' as understood in the HRBAP. Other human rights instruments, for example those relating to the rights of women and of children, set down specific requirements for the consultation and participation of these groups in certain decision-making processes, which is certainly closer to what is intended in the HRBAP. There are also various civil rights – freedom of expression and information, freedom of association and assembly – that are relevant to the capacity of people to participate in social and political activity. These are important rights, but it is debatable whether they constitute a 'principle' of participation, at least in the sense understood.

Accountability and the rule of law – again wholly admirable tenets – cannot be accurately described as human rights principles either. Both are far broader principles of law and governance that are, in a sense, hierarchically superior to human rights in that protection of human rights is dependent upon adherence to these principles. It is not being disputed that all of the above are important considerations to be borne in mind in emergency programming. Yet, pointing out that some of them cannot be correctly described as human rights principles is not mere pedantry. If there is to be a distinctive human rights approach, it cannot be simply based upon relabelling all that is good and desirable as 'human rights'.

The claim–duty nexus in the HRBAP is neither an accurate description of the real world nor a correct reflection of human rights law:

> Increasingly, people of good will concerned about poverty are drawn into debate about a series of ultimately impossible legal quandaries – right of whom, against whom, remediable how, and so on – and into institutional projects of codification and reporting familiar from other human rights efforts, without evaluating how these might compare with other uses for this talent and these resources (Kennedy, 2002).

Kennedy may overstate the criticism – there are certainly some claims that are justiciable, as we shall see – but the crucial point is that these purportedly enforceable claims and obligations are not the only way of seeing social and human relationships in the development or emergency context, nor are they necessarily the most useful. However, both Kennedy in his criticism and Jonsson (2005) in his advocacy of a 'pattern analysis' of these claims and duties miss the point: this pattern may not describe the real world, but nor is it contained within human rights doctrine. Teachers and parents are not human rights 'duty bearers' (to cite the example that Jonsson uses to illustrate how this works). Equally, emergency or development professionals are deterred by the assumption that human rights doctrine sees them as 'duty bearers'. Traditionally, only states have human rights obligations. This is a view that has been widely challenged (Alston, 2005; Clapham, 2006), because of the role of two non-state categories in particular – business corporations and armed opposition – but it could not be said with any clarity that obligations have changed at all.

The great advance of rights-based approaches is that they move away from the provision of needs on the basis of charity, and towards the fulfilment of entitlements (Sen, 1999). Their shortcoming, arguably, is that they fail adequately to identify the obstacles to fulfilment of these human rights (Uvin, 2002).

The strength of rights-based approaches, properly understood, is that they return the focus to the obligations of the state in relation to human rights (Mander, 2005; Gready, 2009). As a normative system, human rights are essentially legal in character, containing the idea of enforceability. Some writers on human rights reject this notion, arguing instead that rights are in some sense an expression of the aspirations of the oppressed (Shivji, 1989). Yet, this seems to make human rights bear a weight that they cannot carry and to devalue the actual contribution that human rights may make to development or emergency programming. I understand this contribution to be closely linked to the commitments that states themselves have undertaken, under international law, to respect, protect and fulfil human rights (Said, 2001; Hathaway, 2005).[5]

5. Human rights in disaster response: Two case studies

5.1. Sri Lanka

Sri Lanka after the 2004 tsunami offers an interesting – and so far unique – example of the potential role of human rights in disaster response, including most importantly the provision of shelter. Just 15 days after the tsunami, the Human Rights Commission of Sri Lanka established the Disaster Relief Monitoring Unit 'with a special mandate of monitoring government and non-government activities which would have human rights implications on the lives of the tsunami affected' (DRMU, 2006).

The Human Rights Commission is a national human rights institution (NHRI) established by statute to monitor and promote observance of 'fundamental rights' and to 'make recommendations to the Government regarding measures which should be taken to ensure that national laws and administrative practices are in accordance with international human rights norms and standards' (HRC Act, 10(d)). NHRIs are an increasingly common phenomenon – there are more than 100 worldwide – as a mechanism for promoting active compliance with international human rights law at the domestic level (Carver, 2010). The Sri Lankan Commission has no direct enforcement powers, which is typical of NHRIs, but is nevertheless generally effective in monitoring government human rights performance and resolving individual complaints from citizens. As a subsidiary body of the Commission, the Disaster Relief Monitoring Unit (DRMU) enjoyed the same powers as the parent institution.

The terms of reference included monitoring government services to tsunami victims, upholding the economic and social rights of victims, monitoring violations of human rights and the rule of law relating to the tsunami, offering recommendations to the government on protecting human rights in the relief and reconstruction process, formulating a code of conduct for government officials engaged in relief and reconstruction, and maintaining liaison with the various governmental and non-governmental agencies. The policy of the DRMU laid particular emphasis on non-discrimination in relief provision, with consideration for the rights of women and children. It also addressed potential friction between those displaced by the tsunami and displaced victims of the continuing conflict in the country (DRMU, 2006).

The DRMU conducted three main activities. First, it conducted regular field visits to monitor the conduct of the relief and reconstruction effort. Second, it set up a programme of consultation with those affected by the tsunami – the assumption being that government policy largely ignored consultation and participation of local communities, particularly the most vulnerable. Thirdly, the DRMU received complaints from affected individuals and communities (DRMU, 2006). By late 2006, the unit had received 10,761 complaints, of which 6,723 (62.5 per cent) related to housing issues (Lyons and Schilderman, 2007).

The shelter programme in post-tsunami Sri Lanka was prey to all the human rights abuses that might be expected – corruption or abuse of power by officials, discrimination in provision of support and the use of the population clearance caused by the disaster to pursue other political or economic objectives. There was also a very strong tendency to ignore the expressed needs of those affected by the tsunami. The complaints process addressed many of the former issues, while the consultation process facilitated by the DRMU allowed better consideration of cultural norms in housing, as well as the particular needs of women, the elderly and people with disabilities (Hidellage, 2008). What distinguished Sri Lanka was not a government with a particularly strong commitment to human rights, but a strong institutional framework that was able to check many serious abuses of rights in post-disaster housing.

5.2. Haiti[6]

All disasters appear to throw up a uniquely complex conjunction of obstacles to relief and reconstruction. Whereas Sri Lanka experienced the tsunami in the context of long-term violent conflict and rural poverty, the Haiti earthquake of January 2010 hit a country marked by chronic institutional instability and extreme urban poverty. Crucially, Haiti lacked any equivalent domestic human rights protection mechanism that could play the role assumed by the Human Right Commission of Sri Lanka. Yet, human rights issues arose just as much in Haiti as they had done in Sri Lanka.

One factor that made the Haitian situation almost uniquely difficult was that the government had itself been to a considerable extent physically destroyed by the earthquake – buildings collapsed and large numbers of government officials died. So the paradox that has been noted – an increase in pressure for accountability at the very moment when the authorities are rendered powerless – was evident to an overwhelming extent in Haiti.

The particular challenge for the construction of post-disaster shelter has been the densely urban character of the affected area. The response, to a large extent, has been the creation of camps, both spontaneous settlements immediately created by earthquake survivors and purpose-built camps further outside Port-au-Prince at Corail and Tabarre Issa. The former, while closer to the city, have very poor conditions. The latter, in better conditions, are far from the means of livelihood that are the priority for survivors of the disaster. A human rights approach would stress the interrelatedness of the rights involved – shelter and livelihoods. In that sense, it would point towards the same solution as that proposed by UN HABITAT, which stresses the need for 'safe return' to neighbourhoods of origin. But it cannot be realistically claimed that it makes the knotty reality any easier to untangle. Human rights principles governing forced evictions would also be important in the execution of any strategy based upon emptying camps.

One area where human rights standards could and should have been applied is in the treatment of women. A Haitian women's rights organization, Solidarité Fenm Ayisiyen, monitored the response of international non-governmental organizations and concluded that many of them 'did not pay adequate attention to women and girls' dignity in camps, especially in the first distributions and in ensuring safe places for women to take showers and to use other facilities. Issues included unrecorded cases of harassment' (Sanderson et al., 2011).

Since it is primarily a law-based normative system, reference to human rights could also have helped to resolve some of the extremely complex issues of land ownership that have arisen. These include (as always) the difficulty of demonstrating title to land or housing, when documents have been destroyed; the construction of camps or transitional shelters on land owned by others; and the rights of squatters, renters or others with no formal title to land. These issues commonly arise in post-disaster situations – as in Sri Lanka – but the urban environment of Port-au-Prince renders some of them particularly intense. A monitoring and complaints mechanism such as that in Sri Lanka might well be helpful, but it may be fanciful to imagine that it could have made a substantial difference.

6. Discussion

There is considerable, and justifiable, scepticism about the introduction of rights language into the arena of development assistance and perhaps even more into the sphere of disaster response. Although 'rights-based approaches' now abound, it is reasonable to ask what human rights brings to disaster response that is both new and relevant. There is a strong sense that rights-based approaches often amount to a repackaging of old practices, a supposition that some international agencies have reinforced. Both the World Bank and UNDP, when adopting HRBAPs, have claimed that their past work was always about human rights anyway – begging the question of why this rhetorical change was necessary (Uvin, 2002).

Some of the elements of a human rights-based approach that are said to be particularly distinctive are not so in reality. 'Participation' is described, not altogether correctly, as a human rights principle and a distinct characteristic of HRBAPs. Yet, participatory approaches to development predate HRBAPs by many years and are to be found in the work of scholars, as well as the practice of many organizations (Chambers, 1983).

When it comes to the practicalities of constructing shelter immediately after disaster, the seven criteria for adequate housing scarcely come into play – the recent example of Haiti illustrates this as clearly as any other. The adoption of a human rights-based approach will not greatly affect the shape of the shelter available. What human rights do bring, however, is an understanding that shelter should be provided on a non-discriminatory basis on the grounds of *entitlement*. Of course, the need for non-discrimination may be well understood whether or not agencies providing shelter after disaster are using an HRBAP. However, human rights make the notions of universal entitlement, equality and non-discrimination integral to any humanitarian programme. Such notions underpinned the successful role of the DRMU in Sri Lanka.

Mention was made above of the integration of humanitarian assistance and protection functions in the IASC guidelines. Protection, broadly speaking, can be understood to mean monitoring and advocating respect for rights. The United Nations High Commissioner for Refugees is a salutary example of what can happen when the protection function – UNHCR's core role under international law – becomes subordinated to the provision and coordination of humanitarian assistance (Goodwin-Gill, 1999; Loescher, 2003). Rather than monitoring and holding states to their obligations to protect refugees, UNHCR has often been party to 'humanitarian' solutions – such as the accommodation of refugees in camps – that have led to serious violations of their rights (Verdirame and Harrell-Bond, 2005).

While human rights have little to say about what shelter after disaster should look like, they do provide a necessary framework for understanding the entitlements of those to whom shelter is provided. In part this means that shelter should be provided purely on the basis of need, but also that this should be done without violating other rights to which the displaced are entitled. This article has quoted the international standards selectively, to isolate what is said on the specific issue of adequate housing. But these standards are best understood as a whole. Human rights are seen as interrelated and interdependent. Hence, it is not acceptable, for example, to make provision of shelter dependent upon the surrender of other rights, such as freedom of movement or certain political freedoms, as is sometimes the case when shelter is provided in camps. Nor is it possible, in human rights terms, to make a trade-off between shelter and other economic rights, such as employment. Although the seven criteria for adequate housing cannot offer strict and enforceable standards in an immediate post-disaster context, they do provide guidance for planning shelter after disaster, particular the intersection of the shelter dimension with non-discrimination and access to services.

Where a human rights approach becomes particularly valuable, however, is in the transition from the provision of shelter as an immediate emergency response into longer-term reconstruction or development. In the first phase, the leeway for choice by agencies is limited and their actions are, almost by definition, responsive. Governments, which bear the primary obligation for protecting human rights, often find themselves close to powerless. Moving beyond shelter after disaster, however, entails clear policy choices by all actors, with national authorities much more likely to make decisions that will affect future enjoyment of human rights.

Most evidently, the seven criteria that comprise the human right to adequate housing can now be realistically integrated into plans for the construction of intermediate or long-term housing for the displaced. Of course, mere invocation of human rights principles does not magically resolve practical problems. The dilemma in Haiti – safe and physically adequate housing or shelter that is close to sources of livelihood – becomes no less acute when it is rearticulated in human rights terms. However, using the language of citizen entitlement may help to focus the attention of policymakers on the nature of their obligations.

Particular mention needs to be made of the first criterion in the CESCR's definition of the right to adequate housing: security of tenure. The UN Special Rapporteur on the Right to Adequate Housing identified security of tenure as one of three 'crucial entry points to incorporate the right to adequate housing in the reconstruction efforts' (the other two being participation and institutional coordination) (Human Rights Council, 2010). This issue arises in several guises in the post-disaster context and, being an essentially legal issue, is one where human rights clearly point the way to solutions. At a minimum, the following issues may arise:

- How can displaced people prove title to property (whether destroyed or abandoned), in situations where documents may be destroyed or there may have been no written title in the first place?

- How can the tenure rights of those whose land is temporarily occupied in the post-disaster phase be guaranteed?
- What procedural or substantive protections exist for those who are removed for their own safety or to facilitate reconstruction?
- What procedure exists for reclaiming property after displacement?

Very importantly, human rights should be able to address the common problem that a simple legal approach cannot: that many people, usually the poorest and most vulnerable, did not enjoy security of tenure, or any legal title, *before* the disaster took place and that the catastrophe serves as a very effective bailiff to evict them. The issue of government failure to resettle displaced people in their original homes has become a thread running through recent disaster responses. In all the countries affected by the 2004 tsunami, governments have prohibited or discouraged local people from rebuilding housing on the coastal strip, ostensibly for their own safety, but accompanied by incentives for the development of tourist resorts in the coastal buffer zone (Klein, 2007). The value of a rights-based response is that even where such measures have been imposed lawfully in terms of local planning law (which is by no means always so) it would be possible to mount an effective challenge on the basis of human rights standards that would be hierarchically superior.

This takes us close to understanding the distinctive contribution that human rights may have to offer in the field of post-disaster shelter. Rights-based approaches to programming reflect norms and principles that may or may not be useful aids to humanitarian practice, but they do so for actors who do not for the most part themselves have hard human rights obligations. In contrast to the UN's HRBAP, which posits a web of claim–duty relationships, human rights law actually creates only one nexus of obligation, which is from the state to the international community. It is simply incorrect (and hence pointless) to say that builders, planners, architects, aid workers and so on have human rights obligations on an individual basis, although this is precisely the assumption underlying the UN HRBAP. There is a general expectation that everyone will adhere to the principles contained in the UDHR, but no enforceable legal obligations flow from this.

In contrast, states do have precise and potentially enforceable obligations under human rights law. In the past decade or so, there has been a widening stream of cases in national and regional courts, as well as treaty monitoring bodies, addressing such issues as the obligations of the state to house the homeless, non-discrimination in housing provision, protection against forced eviction, conditions for eviction for the purpose of developing new housing and the relationship between location of housing and access to livelihoods (Grootboom, 2000; European Roma Rights Centre v Greece, 2003; Selçuk and Asker v Turkey, 1998; Port Elizabeth, 2004; Olga Tellis, 1985). These are all highly relevant issues for post-disaster housing that, if not in practice enforceable at all times and in all places, are at least potentially so.

Of course, there is an irony here, which may be the underlying explanation for the perceived irrelevance of human rights in disaster response: human rights posits a set of obligations on the state at the very moment when the state is, in good faith, unable to fulfil its most basic duties, Haiti being a clear recent example. The danger is that human rights advocates appear idealistic and impractical. Yet, recent experience, for example in several of the countries affected by the 2004 tsunami, suggests that the post-disaster period is precisely when close monitoring of adherence to human rights standards is necessary. Governments use the perception of chaos following disaster to impose solutions – particularly in relation to shelter and property – that are not in compliance with their human rights obligations.

State human rights duties apply in another respect too. These are now commonly understood to extend beyond a state's borders. Hence, there are broad obligations contained within the ICESCR for international assistance to realize economic, social and cultural rights (ICESCR: Article 2). There are also more specific obligations for states to respect and protect human rights in their extraterritorial activities (Coomans and Kamminga, 2004). This implies that governmental (and arguably intergovernmental) agencies involved in humanitarian assistance must comply with human rights standards in their work.

7. Conclusion

In the debate about the relationship of human rights and humanitarianism, the tendency has been to overstate either the potential or the dangers of invoking human rights in the responses to emergencies. On the one side, advocates of 'traditional' humanitarianism fear that human rights politicizes humanitarian action with an emphasis on 'root causes'. This is a curious criticism, given that human rights are equally frequently criticized for their lack of conjunctural specificity and *failure* to attend to root causes (Kennedy, 2002). The human rights doctrine of universality and indivisibility is understood to mean that an HRBAP approach does not allow prioritization between issues, which would be clearly dangerous and impractical in an emergency. Yet, this too misunderstands human rights thinking. Not only does human rights law specifically allow for derogations from certain rights in times of emergency; it also recognizes a hierarchy of rights, with most rights deemed to be legitimately subject to limitation (Seiderman, 2001). Much human rights scholarship is to do with the issue of how to balance competing rights, in theory or practice. That a human rights approach allows prioritization of different aspects of the right to housing at different times should be apparent from the preceding discussion.

On the other hand, supporters of human rights approaches to humanitarianism also tend to overstate their case. There is a tendency to see human rights as a single all-encompassing answer to all questions. Human rights advocates have a tendency to ignore the contribution to be derived from other ways of thinking about the world. Participatory approaches are not the exclusive preserve of the HRBAP. Human rights make a positive contribution when they talk about entitlement rather than need, but they create confusion or worse when they posit a complex network of claims and duties that are unsupported in human rights law or doctrine.

It should be possible for us to hold these two thoughts simultaneously:

- Human rights are applicable at all times, including in emergency, post-disaster situations.
- Human rights do not offer the answer to all programming issues in humanitarian response to disasters.

The debate then becomes not whether human rights are relevant to emergency responses, but *how* in practical terms this might be so. It might then be possible to arrive at a modest approach that might fall short of the full-blown 'human rights based-approach to programming', at least in its UN iteration, but would ensure that human rights are respected, both in the initial emergency response and in longer-term reconstruction of housing.

Acknowledgments

I have benefited from David Sanderson's insights on Haiti and Lisa Handley's unwavering scepticism about the right to adequate housing.

Notes

1. UN treaty monitoring bodies issue General Comments to elaborate particular aspects of the rights contained within the treaties, or to address particular issues of interpretation. These interpretations are regarded as authoritative in explicating the obligations that states incur, although they are not in themselves binding.
2. The Sphere Guidelines grew out of an initiative launched in 1997 by a number of humanitarian organizations, including the Red Cross and Red Crescent movement. They reflect a rights-based

approach to humanitarian relief, as well as providing much more detailed standards for shelter (and other sectors) than are offered by the IASC Guidelines.

3. There is considerable difference between various human rights-based approaches. The UN approaches are discussed here because they are the most widely applied and contain elements that are common to all HRBAPs. However, specific criticism of the UN approach will not necessarily be applicable to the approaches of other international organizations or NGOs.
4. Gould (2009), following Kent (2001), says that arguably there *should* be a 'right to disaster mitigation', but neither claims that such a right actually exists.
5. Hence Edward Said:

   ... [T]o be 'for' human rights means ... to be willing to venture interpretations of those rights in the same place and with the same language employed by the dominant power, to dispute its hierarchy and methods, to elucidate what it has hidden, to pronounce what it has silenced or rendered unpronounceable.

   James Hathaway has a succinct and compelling argument on the actual scope of universal human rights in international law.
6. The discussion in this section draws largely on factual data presented in the report to the Disasters Emergency Committee of a team led by Professor David Sanderson (2011).

References

Advisory Group on Forced Evictions, 2010. *Mission Report to New Orleans, United States*. Advisory Group on Forced Evictions.

Alston, P., 2005. *Non-State Actors and Human Rights*. Collected Courses of the Academy of European Law. Oxford University Press, Oxford, New York, ix, 387.

Carver, R., 2010. A new answer to an old question: national human rights institutions and the domestication of international law. *Human Rights Law Review*, 10(1). 1–32.

Chambers, R., 1983. *Rural Development: Putting the Last First*. Longman, London.

Clapham, A., 2006. *Human Rights Obligations of Non-State Actors*. Collected Courses of the Academy of European Law. Oxford University Press, Oxford, xxxiii, 613p.

Coomans, F. and Kamminga, M. T., 2004. *Extraterritorial Application of Human Rights Treaties*. Intersentia, Antwerp.

Cornwall, A. and Nyamu-Musembi, C., 2004. Putting the 'rights-based approach' to development into perspective. *Third World Quarterly*, 25(8). 1415–1437.

Darcy, J., 2004. Human Rights and Humanitarian Action: a review of the issues. *Human Rights and Humanitarian Action*. Overseas Development Institute, Geneva.

DRMU, 2006. *Annual Report*. Human Rights Commission of Sri Lanka, Colombo.

Goodwin-Gill, G. S., 1999. Refugee identity and protection's fading prospects. *Refugee Rights and Realities: Evolving International Concepts and Regimes*. F. Nicholson and P. Twomey (eds). Cambridge University Press, Cambridge.

Gould, C. W., 2009. The right to housing recovery after natural disasters. *Harvard Human Rights Journal*, 22(2). 36.

Gready, P., 2009. Reasons to be cautious about evidence and evaluation: rights-based approaches to development and the emerging culture of evaluation. *Journal of Human Rights Practice*, 1(3). 380–401.

Gready, P. and Ensor, J., 2005. *Reinventing Development? Translating Rights-Based Approaches from Theory into Practice*. Zed Books, London.

Harrell-Bond, B., 1986. *Imposing Aid*. Oxford University Press, Oxford.

Hathaway, J. C., 2005. *The Rights of Refugees under International Law*. Cambridge University Press, Cambridge.

Hidellage, P. V., 2008. Adapting strategies to meet needs in post-disaster reconstruction. *Building Resilience: Achieving Effective Post-Disaster Reconstruction*. International Group for Research and Information on Post-Disaster Reconstruction, Christchurch, New Zealand, 10.

Housing and Land Rights Network, 2006. *The Human Right to Participation in Post-tsunami Housing Reconstruction: Report of a Factfinding Mission to Tsunami-Affected Areas of Tamil Nadu and Pondicherry*. Housing and Land Rights Network.

Jonsson, U., 2005. A human rights-based approach to programming. *Reinventing Development? Translating Rights-Based Approaches from Theory into Practice*. P. Gready and J. Ensor (eds). Zed Books, London.

Kälin, W., 2000. *Guiding Principles on Internal Displacement: Annotations*. American Society of International Law; Brookings Institution, Washington, DC.

Kennedy, D., 2002. The international human rights movement: part of the problem? *Harvard Human Rights Journal*, 15(Spring). 101–125.

Kent, G., 2001. The human right to disaster mitigation and relief. *Environmental Hazards*, 3. 137–138.

Klein, N., 2007. *The Shock Doctrine*. Penguin, London.

Loescher, G., 2003. UNHCR at fifty: refugee protection and world politics. *Problems of Protection: The UNHCR, Refugees, and Human Rights*. N. Steiner, M. Gibney and G. Loescher (eds). Routledge, New York.

Lyons, M. and Schilderman, T., 2007. Competing approaches to permanent housing in Sri Lanka. Powerpoint presentation, accessed 15 May.

Mander, H., 2005. Rights as struggle – towards a more just and humane world. *Reinventing Development: Translating Rights-Based Approaches from Theory into Practice*. P. Gready and J. Ensor (eds). Zed Books, London.

Phuong, C., 2004. *The International Protection of Internally Displaced Persons*. Cambridge Studies in International and Comparative Law. Cambridge University Press, Cambridge.

Said, E., 2001. *Reflections on Exile and Other Literary and Cultural Essays*. Granta, London.

Sanderson, D. *et al.*, 2011. *Urban Disasters – Lessons from Haiti*. Disasters Emergency Committee, London.

Seiderman, I. D., 2001. *Hierarchy in International Law: The Human Rights Dimension*. Intersentia Hart, Antwerp, Groningen, Oxford.

Sen, A., 1999. *Development as Freedom*. Oxford University Press, Oxford.

Shivji, I. G., 1989. *The Concept of Human Rights in Africa*. CODESRIA, London.

Uvin, P., 2002. On high moral ground: the incorporation of human rights by the development enterprise. *Praxis: The Fletcher Journal of Development Studies*, XVII. 1–11.

Verdirame, G. and Harrell-Bond, B., 2005. *Rights in Exile: Janus-Faced Humanitarianism*. Berghahn, Oxford.

Cases and authorities

European Roma Rights Centre v Greece, 2003. European Committee on Social Rights, Complaint No. 15/2003 (8 December 2004).

General Comment No. 4 'The Right to Adequate Housing' (Art. 11 (1) of the Covenant), Committee on Economic Social and Cultural Rights, 1991.

Grootboom v Oostenberg Municipality et al. Constitutional Court of South Africa, Case CCT 11/00 (4 October 2000).

Human Rights Commission Act (Sri Lanka), Act No. 21 of 1996.

International Covenant on Economic, Social and Cultural Rights, 16 December 1966 (entered into force 3 January 1976).

Office of the United Nations High Commissioner for Human Rights, *Frequently Asked Questions on a Human Rights based Approach to Development Cooperation*.

Olga Tellis v Bombay Municipality Corporation, 1985. 2 Supp SCR 51 (India); (1987) LRC (Const) 351 (Supreme Court of India).

Protecting Persons Affected by Natural Disasters: IASC Operational Guidelines on Human Rights and Natural Disasters, 2006. Brookings Institution, Washington, DC.

Port Elizabeth Municipality v. Various Occupiers, 2004. Constitutional Court, (12) BCLR 1268 (CC).

United Nations Principles on Housing and Property Restitution for Refugees and Displaced Persons (The Pinheiro Principles), 2005. Centre on Housing Rights and Evictions, Geneva.

United Nations, 2006. *The Human Rights based Approach to Development Cooperation: Towards a Common Understanding Among UN Agencies*, 2003.

United Nations Human Rights Council, 2010. *Report of the Special Rapporteur on adequate housing as a component of the right to an adequate standards of living, and on the right to non-discrimination in this context, Raquel Rolnik*, A/HRC/16/42.

Universal Declaration of Human Rights, 10 December 1948.

Unnikrishnan J.P. v. State of Andhra Pradesh, 1993. 1 SCC 645.

Selçuk and Asker v Turkey, 1998. European Court of Human Rights, 26 EHRR 477.

Sphere Handbook 2011, available at www.sphereproject.org/component/option,com_docman/task,cat_view/gid,17/Itemid,203/lang,english/.

Can humanitarian responses in urban areas reinforce underlying causes of vulnerability? Tweaking a livelihoods analysis of inequality and infrastructure in splintering cities

CATHERINE ANNA CRAWFORD[1,2]

[1]CARE International UK, 10-13 Rushworth Street, SE1 0RB, London, UK

[2]University College London, Civil, Environmental and Geomatic Engineering, Chadwick Building, Gower Street, WC1E 6BT, London, UK

International humanitarian agencies are often mandated to target the most vulnerable. This, combined with short funding cycles, encourages extractive assessments of individualized vulnerability; responses based on delivering individual packages of household assets; and only superficial consideration of power relationships. By analogy with urban infrastructure provision, this paper argues that narrow interpretation of the sustainable livelihoods framework applied to urban analysis gives only limited insights into the systemic risks faced by urban dwellers. Using an analysis of infrastructure that draws on the idea of splintering urbanism, a process by which urban infrastructure tends to privilege some and bypass others, the paper develops a visual framework for considering the interdependence of urban livelihoods, physical infrastructure and the organizing tendencies of powerful service providers. The paper concludes that in the cases where it is not possible to identify the most vulnerable (e.g. in dense, heterogeneous cities) or where it is possible only to identify the most vulnerable but not to deliver asset packages to them (e.g. transitional shelter kits that cannot be received by people without access to a plot), humanitarian activities risk reinforcing the systemic, underlying causes of vulnerability.

1. Introduction: Urban inequality and humanitarian response

This paper asks whether the organization of the humanitarian sector has a tendency to reinforce existing inequalities in cities. The emphasis is on urban areas, in recognition of the facts that the number of urban dwellers is expected to increase from 3.3 billion to 5 billion by 2030 (UN-HABITAT, 2006, p. 10); that '[m]any cities are outgrowing the capacity of roads, water supply, and sewage disposal systems to serve their inhabitants' (World Bank, 2010, p. 170); that '[t]he concentration of populations in urban areas will change the nature of many humanitarian disasters… [m]ore people will be living on marginal land, in overcrowded and poorly planned housing, lacking access to adequate water and sanitation, health care and education' (HERR, 2011, p. 9); and that the Inter-Agency Standing Committee Task Force on Meeting Humanitarian Challenges in Urban Areas notes that rapid, ill-managed or uncontrolled urbanization can generate or exacerbate humanitarian crises (IASC, 2009).

Cities already face persistent and stark inequalities in living conditions, tenure security and the geographic distribution of infrastructure (UN-HABITAT 2003, p. 145). Interventions after disasters will be overlaid onto these mosaics such that: 'assistance delivery modalities – whether through the establishment of internally displaced person (IDP) camps, the organization of resettlement schemes, delivering aid to populations in rural areas or supporting the absorption of displaced people into the urban fabric – are an important determinant of whether the affected population loses, keeps or gains access to land and whether people can establish sustainable livelihoods' (Pantuliano, 2009a, p. 201). At the same time, one of the few studies of the economic impact of shelter provision after disasters by CHF for USAID found that 'controlling for factors such as age and vulnerability of recipients, the incomes of persons who were provided with shelter assistance tended to increase at a faster rate than the incomes of persons who were not provided with assistance, and that this effect persisted for years after the provision of the assistance' (Setchell, 2005, p. 1). This highlights the relative, long-term negative impact of missing out on poorly targeted support but it comes with an important caveat pointed out by the Tsunami Evaluation Coalition:

> Many recovery projects focused on physical assets. This meant that those who had held physical assets before had them replaced, with the effect of providing more aid for the better-off than for the poor (Tsunami Evaluation Coalition, 2007, p. 18).

2. Justification for a theoretical approach

This paper takes a theoretical approach to the question of whether humanitarian action might reinforce existing biases. It does this by (a) making an analogy with modes of organizing urban infrastructure, specifically the challenges in delivering water and sanitation across cities; (b) tackling some of the conceptual limitations of using the sustainable livelihoods framework to analyse urban life; (c) offering an extension to one aspect of the livelihoods framework by borrowing from studies of urbanism; and (d) drawing parallels between the imperatives of humanitarian actors and other providers in urban settings and asking whether the urge to identify and quantify urban vulnerability immediately after a disaster overlooks and eventually embeds vulnerability rather than resilience.

The choice to use an analogy with infrastructure provision is prompted by two strands of the current humanitarian debate that echo debates on urban services and will be explored in Section 2. The first is the role of the private sector, which is often advocated with a limited analysis of the capacities, needs and imperatives of this sector (Clermont et al., 2011; HERR, 2011), and the second is the notion, touched upon by Alex de Waal's essay in 'Uncharted Territory', that humanitarian agencies might play a role as 'benign service providers'. Choosing the livelihoods framework reflects, on the one hand, the increasing use of such frameworks by humanitarians, as noted by de Waal (2009, p. 9) and evidenced by the Norwegian Refugee Council's recent guidelines (Suvatne, 2010), and, on the other hand, the criticism of the approaches to livelihoods taken by NGOs that have tended to take a view of livelihoods that is 'stylised and often media-driven' (Tsunami Evaluation Coalition, 2006, p. 29) and that romanticizes 'own-account production as inherently more "sustainable"' in ways that result in 'narrow and inaccurate conceptions of how best to promote equity' (Tsunami Evaluation Coalition 2006, p. 10).

Any attempt to put together the terminology of urbanism, livelihoods and disaster response is likely to be awkward, since it must straddle academic writing, which dwells in critical theory and uses impenetrable language, and practitioner reporting that is all about pragmatism, solutions, regrettable cock-ups and worthy lesson learning. What the different observers share, however, is a lived confrontation with the dilemmas of the built environment and an awareness that the choices we make responding to a disaster or our arguments about why cities are patchy will depend on both theoretical *and* political positions.

In an essay titled 'Space as a Key Word', David Harvey imagines, after Lefebvre, that the real, material stuff in space – the points and lines, movements and flows; relationships and connections – is not only conceptualized, as people try to describe and map the world, but also lived in. Living in the space where the stuff is put together, we find ways to describe what is built; we act and react; and we remake our descriptions of the built environment and continue to build and change it. For those intervening, working quickly to fix and cement spaces that others will have to inhabit, this is potentially troubling. As an example, the author was once asked to review a series of designs for rural schools in Country X and to take into account the specific needs of girls, but – even after conversations with the girls, boys and families – the author was without the language or imagination to know whether the spaces inside the building would make the girls feel safe or trapped; be looked after or spied upon; and condemn them to being *only* inside or stir up revolt. If this isolated school building had been part of a city, the riot of identities, the accidents of encounter and avoidance, the things buried underground, covering the surface, rising into space, marked on maps or just conveying radio waves, would be shaping each person's daily life in unknowable ways. Just as unknowable are the implications of decisions made quickly after a disaster by powerful outsiders unfamiliar with the city as it was, or by powerful insiders unable to imagine how it might have changed. This paper tries to make sense of the geographical unevenness of urban infrastructure *before* a disaster, in terms of the policies, people and organizations that drive it across territory. Then, exploring the reactions that people have to infrastructure, collectively and individually, it takes the livelihoods framework – because it has household experience at its centre and is familiar to the humanitarian sector – and advances it by imagining livelihoods connected to the patchy systems and lived experiences of a city and what this will mean for humanitarian response.

The theoretical arguments are inspired by a study conducted in Cusco, Peru in 2006–2007 (Crawford, 2009; Crawford and Bell, 2011), while the brief substantive arguments, outlined in Section 1, are based on evaluations of disaster responses. The paper uses terminology from the sustainable livelihoods framework, but, to ensure that the resilience and vulnerability of households is not seen exclusively from an environmental perspective, the paper refers to 'livelihoods' and takes a broad interpretation of the vulnerability context as being at risk of economic, environmental and political shocks and stresses, including weather and seismic events, evictions or perceived insecurity of tenure and precarious employment. Household resilience is understood narrowly as the opposite of vulnerability: the ability to buffer theses shocks. Risk reduction is considered a reduction in exposure to any of these shocks and stresses, not only natural phenomena but other events that spell disaster for vulnerable individuals, households and groups. These definitions have been chosen deliberately because it is useful for actors in humanitarian shelter to recognize that physical or environmental hazards should be seen alongside other daily or seasonal vulnerabilities – including political and institutional exclusion – that may continue to discourage people from investing in safer housing and infrastructure promoted through training and technical advice. 'Infrastructure' refers initially to urban water and sanitation systems but is then expanded into a broader conception of socio-technical systems: networks of services, assets and relationships that influence people's daily experience in cities. Power and the powerful are understood as the highly privileged and conditioned actors that have influence over policies, institutions and processes.

3. Policies, institutions and processes in urban water and sanitation provision

In setting up an analogy with urban service provision, it is useful to start with the international policy response to infrastructure provision that has evolved from the World Bank's 1994 (World Bank, 1994) privatization prescription towards an emphasis by 2004 on good governance and a triangular partnership between state, citizen/clients and private, non-governmental or community providers (Nickson, 1997; Estache et al., 2001; Budds and McGranahan, 2003; Leipziger, 2003; World Bank, 2004). This shift in emphasis was largely in recognition of the fact that private sector delivery of water and sanitation infrastructure did not turn out to be the cure-all anticipated by the policy prescription favoured in the early 1990s.

A wealth of research in Latin America drew out two further features of infrastructure governance: the need for participation and the coexistence in cities of multiple models of delivery. Ana Hardoy and Ricardo Schusterman, following research on the privatization of water infrastructure in the informal settlements of Buenos Aires, conclude that the models most likely to succeed over time are those that have incorporated participation mechanisms (A Hardoy and Schusterman, 2000, p. 75). In addition, they cite a need for more open attitudes to 'differentiated' levels of service for different consumers (A Hardoy and Schusterman, 2000, p. 75). In Bolivia, Bustamente argues for support to community-managed systems that are consistently overlooked by existing policies (Bustamente et al., 2004, p. 2). In Venezuela, Caracas' unwieldy water entity, *Hidrocapital*, successfully institutionalized the capacity for participation through community Water Forums (Lacabana, 2008, p. 47). A 2003 publication demonstrated that small, entrepreneurial water providers coexist alongside overstretched utility companies, offering more mobile alternatives such as tanked and bottled water (Solo, 2003, p. 12). These, and other detailed case studies, have contributed to Adriana Allen et al.'s analytical framework that shows that peri-urban governance of water and sanitation is driven by both policy and needs (Allen et al., 2006, p. 339).

There is recognition, then, that certain policy-driven models of urban water governance overlook low-income and peri-urban patches of cities rather than producing equitable or universal infrastructures. The divergence of policy and practice in water provision is by no means unique to these examples and finds parallels in other spheres where the penetration of national policy is partial. John Unruh's essay, for example, describes the legal pluralism in urban land tenure systems that arises after 'population displacement and dislocation' or rapid urbanization and the ways that people shop around for the most pragmatic forums through which to resolve disputes (Unruh, 2009, p. 55). Similarly, Pantuliano observes a 'hybrid of customary and statutory forms' for resolving controversial land disputes after conflict (Pantuliano, 2009b, p. 163). In Colombia's long-running internal displacement crisis, while the government 'fulfils its obligations through legislation, legal recourse, and institutional venues for services, it *denies* its obligations at the same time by narrowly defining the eligible beneficiary group, limiting the attention available, and placing obstacles in the way of claiming rights and services. The result has been a highly developed but ineffective response to displacement' (Fagen et al., 2003, p. 53).

So, even where the policy environment is well developed and documented, coverage can be biased and partial such that access to urban systems and services – when their institutions and infrastructure are geographically close in a city – is mediated by technical and social barriers. This will be examined in Section 4.

4. Livelihoods: A link between infrastructure and vulnerability

A closer look at household responses to patchy infrastructure provision is possible through one interpretation of the sustainable livelihoods framework, shown in Figure 1. This analytical framework has already brought together concepts of household, infrastructure and vulnerability governance. From now on, this paper refers to urban livelihoods, to avoid the ambiguity of applying notions of sustainability to urban life (discussed in Section 6). The concept of livelihoods has been integrated into humanitarian standards and guidelines (Sphere Project, 2004); approaches to assessment and response (OXFAM, 2005, p. 145); attempts to integrate humanitarian responses across sectors (Norwegian Refugee Council, 2008, p. 454); greater recognition of the capacities of disaster-affected people, not just their neediness (IFRC, 1999); and the need for participatory approaches as a complement to extractive and quantitative

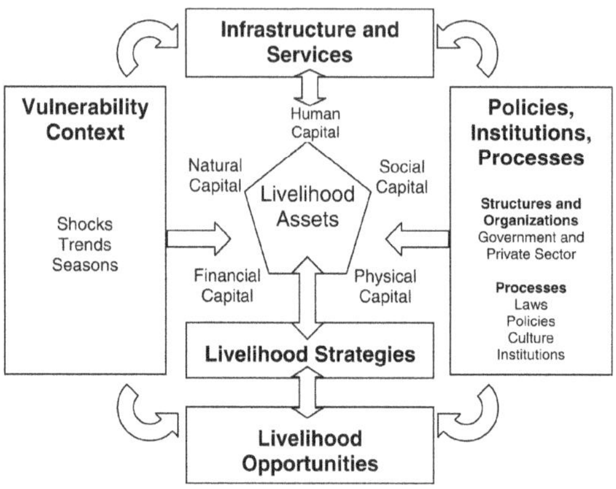

FIGURE 1 The sustainable livelihoods framework adapted from Tony Lloyd-Jones' diagram (Rakodi, 2002a, pp. 8–16).

methods. After a disaster, however, livelihood programmes are typically confined to food security and income-generating activities – the Sphere livelihoods standards appear under food security, indicated by primary production, income and employment and access to markets (Sphere Project, 2011, p. 145) – and there can be a limited understanding among agencies of how different groups assemble their livelihood strategies (Tsunami Evaluation Coalition, 2006, p. 10).

The framework's first analytical category – policies, institutions and processes – captures the fact that, although household activities are at the heart of the framework, individual household livelihood strategies do not operate in a political vacuum (de Haan, 2002; Farrington et al., 2002; Rakodi, 2002b). The penetration of national policy depends on the capacity and will of local authorities to implement and on the ways in which infrastructure is governed locally and historically, be it by committee, community or municipality (Crawford and Bell, 2011). The role of local government in negotiating the formality of settlements, issuing building permits, collecting and spending tax revenue and tussling with national government cannot be overstated but will not be dealt with in detail in this paper.

The second analytical category of the framework deals with infrastructure and services. By treating these as collective assets beyond individual households, the livelihoods framework acknowledges that what happens to be lying around, buried in the ground or hovering in the social ether near households is not automatically accessible to householders. Instead, the framework makes an implicit distinction between assets at the level of the household, which depend on the quality of the connection, and infrastructure and services, which depend on the quality of the networks of markets, services, infrastructures, ecosystems or institutions to which households can potentially connect themselves (Montgomery et al., 2004, p. 31).

The final analytical categories of the framework are (a) the pentagon of assets that householders manage in order to mitigate risk, (b) the vulnerability context or the extent to which individual households are able to manage livelihood strategies to mitigate risk given their access to assets, connections to infrastructure and the prevailing governance arrangements for the city and its systems.

The asset bundle is conceived as a collection of physical, human, social, natural and financial capital. This obviously does not fully describe how households operate in every context, but it offers a simple model, accessible across economic and social disciplines, as a starting point for analysis.

This paper focuses on the infrastructure and services component of the livelihoods framework and its interaction with household vulnerability, but this relationship is only one of many that affect urban livelihood strategies and it is important to elaborate briefly on the role of shelter and land. Shelter, depending on whether it is conceived as a physical structure, a social space or a speculative investment, spans different capital assets at different times and this is one way of looking at what spurs people to invest in housing relative to other household or survival priorities, at the time and incremental stages of the shelter process and at the risk presented by poor housing relative to other risks, such as eviction or insecure access to land. Indeed, as de Waal points out: '[i]ncreasingly, humanitarians have come to use livelihoods frameworks for understanding and designing their interventions and, to the extent that they are concerned with livelihoods, they must be concerned with land' (de Waal, 2009, p. 9).

Security of tenure – or secure access to housing and land assets – plays a role in stabilizing urban livelihoods. Obtaining formal title, for example, is often claimed as the key driver for poverty reduction, improved housing and the extension of service provision into informal settlements but the reality is that a variety of tenure options exist in cities and the influence of land access on vulnerability relative to other household assets and activities is highly variable. In concluding a recent review of the impact of titling programmes, Geoffrey Payne, Alain Durand-Lasserve and Carole Rakodi argue that: 'home ownership is not appropriate for all income groups. This applies particularly to poor and very poor households, who lack adequate incomes to meet long-term financial commitments and need to be able to respond quickly and easily to changing livelihood opportunities. For them, short-term tenure options are preferable, such as rental accommodation. In addition, land titling has not been shown to achieve the social and economic benefits claimed for it; in fact, many of the advantages for which titles are promoted, such as stimulating investment in property improvements, have been realized by less formal increases in tenure status' (Payne et al., 2009, p. 459). This last observation is also supported by CHF's study of post-disaster shelter (Sheppard and Hill, 2005, p. 24).

On the supply side, providing rental accommodation is often an important livelihood strategy for poor urban dwellers and Sunil Kumar argues that viewing the landlord–tenant relationship as exploitative is unhelpful and that the: 'traditional tenure classification of ownership and renting is limited to the final product – the dwelling – and does not encompass the social processes embedded in the interconnected relationships between the tenures and the agents and institutions which delineate frameworks of production and consumption' (Kumar, 2001, p. 2). Relative land-related vulnerabilities faced by different groups definitely deserve scrutiny, however. In case studies in Senegal and South Africa, where residents in informal settlements already enjoyed de facto tenure security, titling had 'a very positive impact on increasing tenure security for women by specifying them on ownership records', but 'the situation for tenants is less positive in that rents in the Senegal case increased sharply as newly regularized shack owners transferred onto them the costs of tenure regularization and physical upgrading' (Payne et al., 2009, p. 447).

Payne et al. also found no clear evidence for a link between formal title and urban service provision. In Colombia, for example, infrastructure providers were mandated to provide services regardless of tenure status; in Mexico and Egypt, titling programmes were not accompanied by any infrastructure upgrades; and in India, 'investments generated by improved security of tenure in the case of Indian cities may create new problems regarding planning and the provision of infrastructure in very densely populated settlements' (Payne et al., 2009, p. 456). The authors also noted that the provision of urban services to newly formalized settlements may be slowed by the fact that many of the areas without formal title also tend to be the places that are environmentally vulnerable or difficult to access and where the cost of installing services is seen to be prohibitive. Tenure arrangements do play a critical role in patchiness but this role depends on location; on actual, relative and perceived risks of informality and on the vagaries of a household's ability to pay rent and the attendant anxieties of tenants and landlords.

The livelihoods framework tries to capture different facets of insecurity through the 'vulnerability context', which acknowledges a spectrum of deprivation that is both temporal and chronic (Ray, 1998, p. 252); is linked to the environment (J Hardoy et al., 2001, p. 346); affects access to resources and influence; and even affects the ability to aspire to influence (Wratten, 1995, p. 14). On this last point, the World Bank now regards these differences or inequalities in access as not just of intrinsic interest but as factors that in themselves act 'to curtail economic dynamism' and 'combine, interact, and are reproduced through interlinked economic, political, and sociocultural processes' (World Bank, 2006, p. 28).

Humanitarian agencies, thus, find that vulnerability is complex and unclear and hard to reconcile their mandates to target (quickly) the most vulnerable or specific groups, such as refugees or IDPs. As a reaction to this, there have been attempts to profile urban populations and their vulnerabilities in more detail. Internal Displacement Monitoring Centre joint research with Tufts University on displacement to urban areas, used household surveys to unpick the 'experience of displacement' and found in some cases that differences between recently arrived IDPs and the urban poor were subtle with location sometimes more important in vulnerability than being an IDP (Jacobsen, 2008). A 2009 study in Johannesburg compared long-term South African residents, internal migrants and foreign-born residents, concluding that there are '[s]pecific forms, indicators and factors of vulnerability faced by residents in several different neighbourhoods' and that vulnerability is 'experienced differently from one area to another and among different population groups' (Misago et al., 2009). Despite this evidence that urban vulnerability is specific and intimately linked to location and livelihood, the tendency of humanitarian actors is to revert to classic vulnerability criteria that focus on individuals (and their presence or absence in households) and ignore the specificities of location and all other analytical categories constructed in the livelihoods framework. In part, this is because: '[i]mmediately after a disaster when the entire support system is unable to cope, all families may be equally vulnerable in terms of their basic survival needs of food, water, medicines, etc. However, that does not mean that once the 'saving lives' phase is over, and slowly the support system begins to recover from the shock, all individuals have the same level of power, identity, connections and resources' (Bhattacharjee et al., 2007, p. 24).

There is a tension, however, between the humanitarian agencies' need to delve deep, assess and profile livelihoods and the fact that singling out the vulnerable can also be intrusive. For Alex de Waal's this '[a]dministrative intrusion places humanitarian agencies in a particularly awkward position. The spirit of humanitarianism is to support and enfranchise the marginalised, dispossessed and weak. One of the most effective weapons of the weak is to remain elusive to authority – to refuse to be counted, to escape formal administration and to keep open clandestine options for livelihood or migration. Yet technically effective relief programmes demand that the subject population is measured, monitored and governed' (de Waal, 2009, p. 21). De Waal goes on to suggest, optimistically, that '[b]y [their] very existence [humanitarian agencies]... demonstrate the possibility of a rule-governed, professional and benign service provider. This function can be extended into areas that are both more contentious and, in the long term, more significant – such as land allocation and the codification of tenure rights' (de Waal, 2009, p. 21). The model of the 'benign service provider' warrants closer inspection and it is here that the analogy with urban service provision becomes useful.

5. Risk, resilience and relationships: Patchy, complex links between livelihoods

5.1. Privilege and bypass

Concepts to help untangle biases in access to urban infrastructure systems are emerging from the discourse on cities and globalization; both the role that cities play in the global economy – as hubs of activity – and the effect of global economic trends on the economic polarization and segregation within them (Borja and Castells, 1997; Sassen, 2001; Montgomery et al., 2004).

This link between local exclusion and global connection is also picked up in Simon Marvin and Stephen Graham's wide-ranging examination of urbanism, where they make a case for the 'splintering' of cities: a process of 'bypass', whereby the strengthening of connections between the most valued people and places – those judged to be lucrative and profitable to the infrastructure service providers – leads to a weakening of the connections between the least valued users and places (Marvin and Graham, 2001, p. 288). They argue not only that the configuration of networked infrastructure is highly biased but that it successively reinforces existing biases. The work draws on a theoretical approach which sees 'modern urbanism as a socio-technical process' (Graham, 2000, p. 115) and the city 'as a process of flows', 'technological networks' and 'a "hybrid" of the natural and cultural, the environmental and the social' (Kaika and Swyngedouw, 2000, p. 121).

The splintering processes of privilege and bypass are driven in part by the global economic forces described by Borja, Castells and Sassen, but also by an interaction between people and technology.

5.2. Local resistance

The view that globalization is the key determinant in shaping cities has been criticized for overlooking the influence of local governance. Richard Batley, for example, suggests that the destiny of cities 'is not determined just at the global level, or by their inherited starting point: local agents can also be influential'. For Batley, Latin America's distinctive urban inequalities are 'products of policy, not of globalisation' and he singles out residential segregation as something that has become an 'administratively consolidated fact' in some Latin American countries driven by tight municipal boundaries and a lack of cross subsidy (Batley, 1997, pp. 338–341). Montgomery et al. also argue that this can enforce 'homogeneity in local social network ties and resources, suppressing some of the diversity in social relations that can benefit the poor' (Montgomery et al., 2004, p. 43).

The prospects for local agency, however, remain uncertain, manifesting themselves in the disobliging messiness of co-existing attempts at universal provision, urban splintering and community management or infrastructure.

5.3. Networking between socio-technical scales

As well as local political resistance to splintering, Marvin and Graham insist that 'perfect splintering' is ultimately scuppered by the technical impossibility of insulating privileged enclaves from urban or regional services such as power and water, concluding with a plea for the development of more modest, local networks and for the exposure and regulation of inequalities in infrastructure connections, perhaps lest the modest and local encourage insulation.

Bas van Vliet, Heather Chappells and Elizabeth Shove examine the technical barriers to insulation in their work *Infrastructures of Consumption*. These authors suggest that: 'Although the general trend is assumed to be one of a shift [away] from a "universal" mode of provision, the reality is a more complex situation in which private and public priorities coexist and in which networks are both converging globally and fragmenting locally' (van Vliet et al., 2005, p. 38).

This team goes on to devise a framework to describe four coexisting, but not necessarily sequential, phases of infrastructure development, considering the technical systems *and* the experiences of users of these systems. More interestingly, they give an idea of configurations of infrastructure that fall between privilege and bypass.

Bringing these ideas together, we start to see that patchy provision – where infrastructure systems are conceived as socio-technical systems – is configured by global processes, local coalitions and sometimes by modes of organization which find unconventional routes or subvert idealistic governance structures. This goes beyond considering infrastructure as physical pieces or lines on a map to the relationships between providers and consumers and 'modes of organization' that respond to the experience, demands or exclusion of certain people. Privilege, bypass and splintering can be drawn on a map but they are also experienced by people and change the ways that people arrive in, walk through and dwell anxiously or confidently in a city.

6. Moving from riskiness to resilience

The Tsunami Evaluation Coalition noted that attempts to support livelihoods after this disaster failed, in part, because the intricacies of individual household livelihoods were not well understood (Tsunami Evaluation Coalition, 2006, p. 29) and, in part, because prejudiced judgements about good, sustainable livelihoods were confused with notions of self-sufficiency that took no account of bigger systems and deeper equity issues (Tsunami Evaluation Coalition, 2006, p. 10). An equitable humanitarian response, then, needs to address the conceptual problems posed by the term 'sustainable livelihood'.

The livelihoods framework emerged from rural research and the possibility of a sustainable urban livelihood has been challenged. A number of authors observe that the notion of environmental sustainability applied to cities conflates environmental impact, which tends to be small for low-income urban dwellers (J Hardoy et al., 2001, p. 346), with environmental health problems (Newman 2006, p. 280): a confusion that risks marginalizing 'the primary environmental concerns of the poor' (McGranahan et al., 1999, p. 108). Further, the blithe labelling of cities as sustainable or unsustainable can mask 'urban and environmental processes that negatively affect some social groups while benefiting others' and those concerned with equity 'always needs to consider the question of who gains and who pays and to ask serious questions about the multiple power relations' (Swyngedouw and Heynen, 2003, p. 901). Robert Chambers, who insisted in 1995 that development professionals should strive to avoid practices that entrench sustainable privilege for 'us' rather than sustainable wellbeing for 'them' (Chambers, 1995, p. 204), reflected 10 years later that: 'the concept of sustainable livelihoods has...been applied selectively only to the poor'. He argues instead for sustainable ways of life that enhance rather than damage the livelihoods of others 'whether through claims, access, international trade, environmental effects or in other ways, now and/or in the future. This, of course, applies most to the rich and powerful, to their lifestyles, and to their impacts both on the livelihoods of others and on the environment through pollution, climate change and resource depletion' (Chambers, 2005, p. 202). The interconnection that Robert Chambers sees between the livelihoods of groups, which he essentially polarizes in much of his writing into the powerful and the disempowered, is of particular importance with respect to shared and geographically stretched resources such as infrastructure and ecosystems where it is '"policies, institutions and processes" which mediate access to the resources and assets required to sustain a livelihood' (de Haan 2002, p. 4).

Robert Chambers sees one way of reducing vulnerability as the adoption of a diverse range of livelihoods, accompanied by complex adaptive responses where 'diversity spreads risks by adding species, enterprises, linkages and activities; complexity reduces risks through redundancy' (Chambers, 1997, p. 172) . Expanding for a moment on this idea, it is useful to look at a simple model of social insurance which distinguishes between two sorts of independent risk; a localized, 'idiosyncratic' risk that might affect a few people and an 'aggregate' risk that affects everybody in the same way. To optimize mutual insurance, households and communities might choose strategies that mitigate exposure to the same set of aggregate shocks. For instance, if all members of a family participate in a similar activity and this principal economic activity is disrupted for some reason, every family member is hit in a similar way such that economists would expect families to 'attempt to diversify their activities or spatial locations to lower the correlation between the incomes of family members'. Obviously, with a large enough group, all uncertainty is theoretically idiosyncratic and independent and smoothing of income fluctuation can be achieved, but the larger the group, the worse the flow and reliability of information and the harder it is to verify incidents and avoid moral hazard: as Ray puts it, an 'informational barrier kicks in and precludes insurance over very large anonymous groups or spatial distance' (Ray, 1998, pp. 597–605).

Chambers and Ray both suggest that households with 'diversity' are better able to buffer risk, a concept that Ray describes as diversity in activity and diversity in geographic reach. Chambers also calls for complexity in livelihoods to reduce risk. Complexity, for Chambers, can be conceived as multiple, occasionally unnecessary, connections that create redundancy such that if a connection is broken there are backup options, and some learning or adaptation can take place over time.

To relate diversity and complexity to notions of local and global, let us argue that the concept of a local group is analogous to a group without complexity and diversity: it experiences risk in the aggregate. The concept of a global group, by contrast, is a group that has diversity or complexity or both and can buffer risks, which it either experiences as idiosyncratic risks or as failures which it can tolerate through backup connections.

The success of diversity, as an objective for urban communities that are to be resilient in the face of risks, depends on households in the same neighbourhoods being exposed to risks that are idiosyncratic or independent of each other but, in fact, seasonal, local and global vulnerabilities are interrelated. This highlights the fine line between heterogeneity per se – which could also be read as development constraining inequality of the sort described, for example, by the World Bank and Ellen Wratten – and heterogeneity in vulnerability – which sees diversity as positive because it means that not everyone will be knocked by the same problems at the same time.

The concept of connection is critical to the socio-technical systems discourse and is raised by Borja and Castells' concept of urban hubs; Marvin and Graham's splintering process; and Chambers' preoccupation with the tendency for power to reinforce itself. In all these ideas there is a possibility for the well-connected preferentially to connect (or preferentially to be connected by service providers) with those that are already well connected, while the least connected remain much less well connected. This configuration, sometimes described as clustered or non-homogenous, is an abstraction of a group in which a few members are much more connected than the rest. For urban hubs and privileged splinters to connect preferentially to each other or to be connected preferentially by service providers, the connection process must be in response to some quality or perceived quality of the hub. In its simplest form this quality might be the number of connections it already has: in other words, having many connections makes the hub attractive (Albert et al., 2000). In a much more complicated socio-technical system made up of livelihood asset bundles, infrastructure connections and the potential to influence governance, this paper argues that attractiveness is related to the perceived riskiness of the household or system to be connected.

This riskiness is often associated with informality of tenure or physical risks such as settlements on steep slopes or in flood-prone zones and, as Payne et al. note citing Bromley, it is not just attributed to individual households but to entire neighbourhoods: '[t]he best house in a bad neighbourhood… is continually burdened by its surroundings…'. This becomes self-reinforcing because in the absence of information, it is safer to assume riskiness than not. Where there is no knowledge about the bypassed, the tendency is to avoid connecting. The well connected, however, produce information about themselves through the fact of being connected – household bills and a history of consumption are generated. Similarly, if this information is to be collected and cross-referenced as proof of occupation to strengthen a tenure claim, for example after a disaster, the connected will have more evidence to hand.

The final points to make here are that (a) the more knowledge service providers have about those they serve, the more sophisticated their provider's response to risk, including the ability to forecast and pre-empt changes in the future; (b) the embedded socio-technical configurations of infrastructure and livelihoods that emerge from these responses will, to some extent, reflect past risks; (c) thus the sustained ability to buffer new risks in the future will probably require some sort of reconfiguration; (d) if livelihoods are to enhance rather than damage the livelihoods of others, this places even more onerous requirements on a system that balances risk: there is a need to mutualize risk in new ways.

7. Combining frameworks: Livelihoods, infrastructure and governance

The key proposition of this paper is that these abstract concepts can be unified within a socio-technical systems perspective constructed around two axes (Figure 2). The horizontal axis is labelled 'what the powerful know' and it represents the degree to which the risks faced by individuals have been correctly quantified by the most powerful service providers, governments and consumers: at one end, the most powerful know everything and, at the other, they know nothing and in between they make assumptions that become more conservative when less is known. At the right-hand extreme, risk cannot be quantified: the right-hand space is not complex. At the left-hand extreme, risk can be quantified and the response is to buffer risks by forming hubs or clusters: the left-hand space is complex.

The vertical axis is labelled 'how the risk is shared' and is one way of looking at the degree to which one livelihood might damage or enhance another: from risk that is individualized to risk that is mutualized. At the bottom end of the axis, risk is buffered only by a single individual with limited capacity to mitigate risk: without much diversity, each person is more vulnerable. Moving up the axis, however, individuals coalesce into households (with livelihood strategies), settlements connected by shared infrastructure and institutions and cities (within a context of shared policies, institutions and processes). The risk sharing space can be conceived as not diverse (bottom) and diverse (top). Where risk sharing is neither diverse nor complex, livelihoods are vulnerable (bottom right quadrant).

The axes intersect at a hypothetical point where the risk sharing group moves from only being able to buffer local, idiosyncratic risks to being able to mitigate aggregate risks. In the terminology of the previous section this can be viewed as a shift from a local group (a population with everyone vulnerable to the same risks) to a global group (a population with different people vulnerable to different risks).

This is an abstract way of conceiving the links between risk sharing, modes of organizing collective infrastructure and governance: it acknowledges the vulnerability context of the livelihoods framework and, at the same time, the potential power imbalance between providers and consumers of infrastructure and services, where power is understood as having both access to information and the capacity to transform it into useful knowledge for making decisions.

In conceptual terms, the argument is that infrastructures mediate between household vulnerability and governance by setting up the group through which risk is shared. The boundaries of this group are drawn both socially, through a negotiated agreement or refusal to share risk or through unwitting and involuntary sharing, and technically, through the quality and extent of infrastructure.

This can also be explored with reference to van Vliet et al.'s modes of organization. Each mode can be placed in its hypothetical location in the framework as shown in Figure 3 and described in Table 1. In addition, the framework also allows for hybrid modes: the coexistence of many different modes across the same space and the possibility of moving between spaces.

At the bottom of the framework, whatever the mode of infrastructure organization, individuals without access to sufficiently diverse and complex asset portfolios are extremely vulnerable,

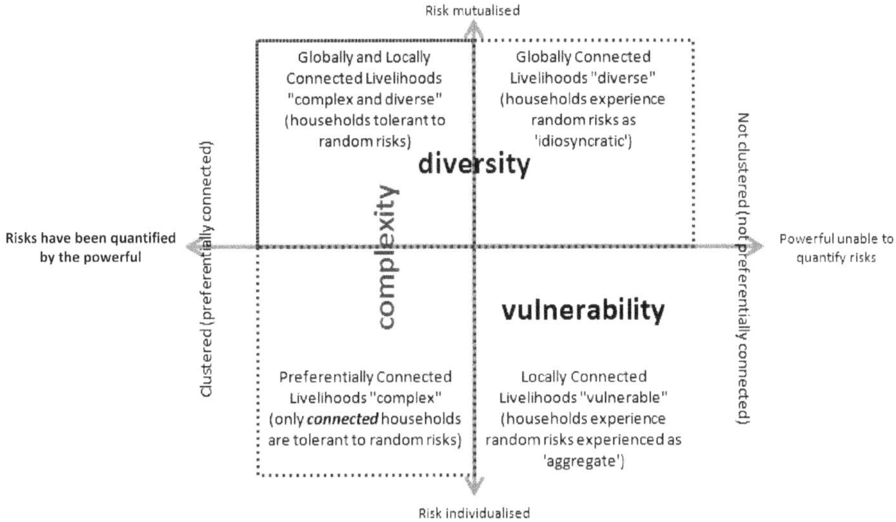

FIGURE 2 Socio-technical conceptual space: vulnerability, infrastructure and governance.

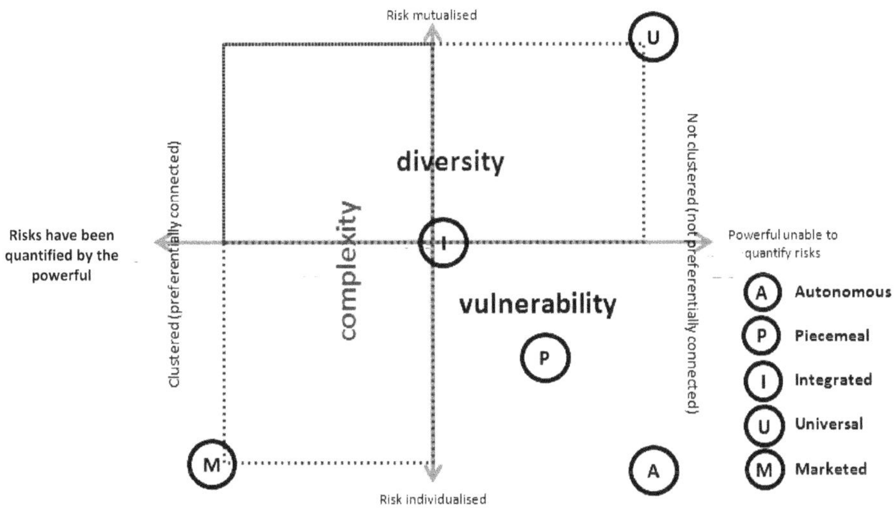

FIGURE 3 Socio-technical conceptual space: plotting van Vliet's modes of organization.

while individuals with sufficiently complex asset portfolios are in a position where they have no need to share risk with the vulnerable and pay only to mitigate their own risk.

At one exaggerated extreme, a provider knows exactly how much each household is able to pay for water, how timely payment will be and exactly when and how much water will be consumed. Pricing can be highly differentiated to optimize cost recovery across a network, with risk (or costs) worked out for each person and, potentially, with the riskiest individuals paying more. This system would be placed in the bottom left quadrant of the plot (M).

The idea of differentiated service provision was posited by Ana Hardoy. For Hardoy, faced with monolithic concessions that cannot accommodate the needs of low-income urban settlements,

TABLE 1 van Vliet's modes of organization reproduced from Table 3.1 (van Vliet et al., 2005, p. 33)

Autonomous	Co-providers, localized resources, local stand-alone, self-managed and responsive grids
Piecemeal	Customers and suppliers, patchwork of grids, unregulated, non-standard, designed for peak loads with idle spare capacity
Integrated	Consumers and promoters of diversified demand, semi-integrated local and regional grids, designed for diverse loads to exploit spare capacity and smooth demand profiles
Universal	Passive beneficiaries and public providers, uniform services, highly integrated national and regional supergrids, demand not differentiated, just met by extending networks
Marketed	Purchasers and promoters of differentiated products and services, partially fragmented grids matched to diverse, monitored and manipulated needs

a differentiated offering is appealing because it suggests that services can be tailored so that the least privileged can choose a service package that fits with their livelihood strategies, just as a range of tenure options was posited by Kumar to cater for those on precarious incomes or undertaking solo, seasonal migration. In the work of van Vliet et al. differentiation is part of a sophisticated provision model that adapts networked services to the highly differentiated demands of users: in this case tailored to lifestyle. The interaction of livelihoods and infrastructure has, therefore, to be carefully understood before differentiated services move from being the observation of a phenomenon to a policy prescription that builds configurations of infrastructure that entrench and undermine livelihoods.

Where a service provider cannot track any of these things, risk (or costs) will be averaged out or shared across the system with a bit of cross-subsidy here and there. This moves us up into the top right quadrant of the plot (U).

Critically, this framework also permits a re-examination of the socio-technical systems perspective in terms of vulnerability and governance. The processes of privilege and bypass described in Marvin and Graham's splintering urbanism thesis can be hypothesized in relation to the framework as shown in Figure 4.

The process of privileging is part of a shift towards a marketed mode of organization in the bottom left quadrant, with powerful service providers motivated (arguably by powerful consumers) to apportion provision and costs according to the risks presented by individual livelihoods. Although risk is individualized for everyone, the privileged, understood as people with complex (interconnected) livelihood assets, are still able to buffer risk. Meanwhile, as the process of privilege consolidates itself, the groups that are left without services find themselves with modes of organization in the bottom right quadrant, bypassed by the socio-technical systems that would allow them to buffer risk. Service providers continue to collect information, and information about privileged users becomes continuously better than information about the bypassed; service providers continue to quantify risk based on this information; thus the riskiness is higher for those already bypassed or in places where lack of information makes risk harder to quantify.

The key point here is the interaction of livelihood assets and vulnerability. The diverse and complex livelihood portfolios that allow the privileged to buffer risk or that tend to make the bypassed vulnerable cannot be understood as something contained within an individual or a household as the livelihoods framework starts to suggest. Diversity is seen, instead, as a stretched socio-technical system that connects a household's assets to other places, while complexity is seen as the link between household assets and other highly interconnected livelihoods, pieces of infrastructure or institutions.

A socio-technical systems perspective thus allows the livelihoods of the privileged and the bypassed to be considered within the same

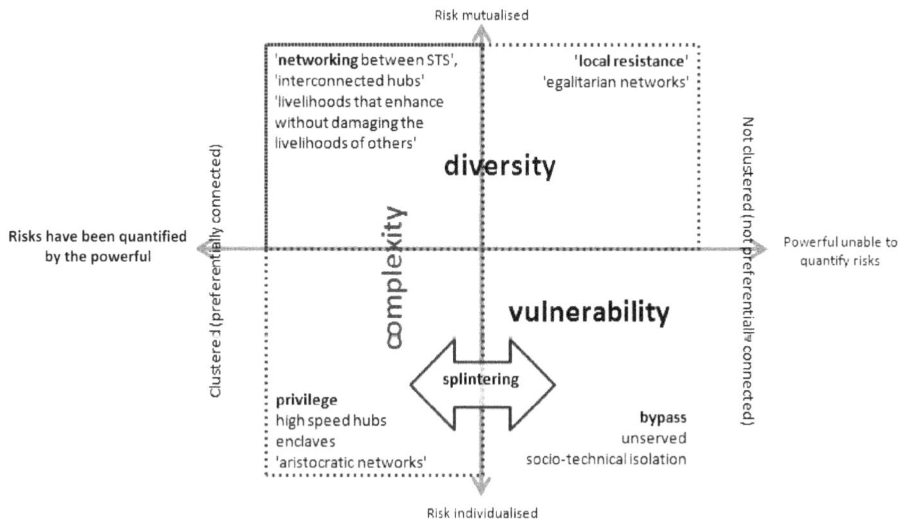

FIGURE 4 Socio-technical conceptual space: plotting Marvin and Graham's splintering.

analysis. This is essential if we are to respond to Chambers' call for sustainable livelihoods that enhance rather than damage the livelihoods of others.

The movement between different quadrants of the framework is driven by models of governance that privilege and bypass on the basis of risks and returns, responding to perceived vulnerability and livelihoods; generate local resistance to bypass by responding to perceived heterogeneity in access to assets; and produce institutional arrangements that can network to the powerful, creating socio-technical systems capable of linking households, committees or municipalities to useful sources of knowledge and finance that extend much further than physical pipes and reservoirs. The quadrants can also be related to humanitarian agencies: where asset replacement is targeted at those who previously held assets, the mode falls into the bottom left quadrant. Where an asset package cannot be received by the most vulnerable, for example when those without access to space cannot receive transitional shelters, the mode of organization tends to privilege those with land access and bypass those without, splintering between the bottom quadrants. Where individual asset packages fail to offer a viable transition, it might be because early recovery activity prioritizes the upgrading of shelters – ignoring the fact that only the select few that have received a shelter package will get an upgrade – or when there is no nuanced support for different livelihood strategies that allow people to connect to markets and suppliers, the overlooked find themselves in the bottom right quadrant.

As with Chambers' positive ideas about diversity, the complex interactions that he thinks can combine to enhance livelihoods are also confounded by the reaction of service providers to riskiness. Generally, where risk is not known it is prejudged, assumed and even estimated forward by financial markets. It is partly this aspect of private sector involvement in infrastructure that picks out safe or profitable nodes, that underlies Estache et al.'s 'secondary distributional effects' (Estache et al., 2001, p. 1180) and shapes the economic, political and sociocultural processes of the World Bank which cause inequalities to 'combine, interact' and reproduce themselves (World Bank, 2006, p. 28). Although humanitarian agencies may not consciously pick out low-risk clients – indeed they are mandated to do just the opposite – strategic decisions to opt for scale rather than impact, logistical or time constraints and coordination failures or decisions to work in convenient rather than needy locations can result in modes of organization that take a similar shape. In Haiti, the shelter cluster's strategy proposed a broad palette of interventions but most agencies moved quickly to large-scale, international procurement of transitional shelter which worked well on flat, open space where agencies could feel confident about security of tenure. This also meant an eventual over-supply of this solution to certain locations: coincidentally flatter, more open spaces (Haiti Shelter Cluster, 2010, p. 11). Other activities, such as repairing housing in densely inhabited areas of the capital, were neglected for two reasons. Firstly, because, from the perspective of engineers, it was more risky to make flawed, damaged housing safe than to build something safe from scratch. Secondly, for NGOs mandated to target the vulnerable, the perception was that fixing damaged rental housing would benefit landlords rather than vulnerable groups such as tenants.

However, both investment in urban infrastructure and urban humanitarian response are interesting spaces for discrimination on the grounds of risk to unfold because, as Richard Batley observes, the dynamic and marginalized are juxtaposed (Batley, 1997, p. 339). In this context, the injustice of privilege and bypass, with respect to damaging rather than enhancing the livelihoods of the bypassed, is often overt and there must be optimism here about the possible collective reaction to risks that affect the livelihoods of the rich and the poor. Secondly, those affected by bypass and privilege in cities are both proximate and numerous. From here, Batley asserts that local agents can also drive change (Batley, 1997, p. 341).

When Alex de Waal suggests that there may be the potential for humanitarian agencies to move from delivering relief to demonstrating 'rule-governed, professional and benign' service provision, his twin assumptions are that 'administrative intrusion' can successfully measure and monitor urban vulnerability and that, having done this, benign service provision will emerge from mandates. This paper argues, by analogy with other forms of urban service provision, that risk aversion begins before de Waal's 'administrative intrusion' and that service provision tends towards splintering modes because:

- administrative intrusion becomes a prerequisite for responding because it provides evidence for strategic and targeting decisions and frames each organization's perception of their internal capacity and impending programme risks. People or places that have been less administratively intruded upon are bypassed perhaps because they are difficult to access (steep, blocked by rubble, served by unsealed roads, inundated or remote); reputed to be unsafe (no-go areas, reported to have a gang presence); or disconnected (without organizations with links beyond the neighbourhood);
- the 'rules' governing service provision are driven by previous experience, commitments to and the expectations of donors and an evaluation of programme risks, usually dominated by a preoccupation with spending on time. Service provision takes the form of asset package distribution and at a scale which requires design and procurement contracts to be set before vulnerability has been understood such that distributions privilege people who can easily receive and make use of a packaged asset, regardless of whether it meets their immediate needs.

There are, then, tendencies that drive powerful providers – governments, civil society and humanitarian actors – into the bottom left quadrant of the model where private sector providers are already ensconced. As humanitarian agencies are being asked to demonstrate value for money, define their limits of operation and coordinate roles and responsibilities with the private sector, there is a need to consider the idea that, in spite of their mandates, the imperatives and modes of organization that drive response may splinter cities in the same ways as those of other powerful actors: privileging the privileged and bypassing the bypassed.

8. Conclusion

This paper asks whether humanitarian modes of operation have a tendency to reinforce existing inequality and then uses a four-pronged argument to answer the question. Firstly, the paper makes an analogy with urban infrastructure, citing evidence that privatization alone cannot achieve full coverage; that consumers engaged in service provision produced better services; and that the reality is that all sorts of community-based and entrepreneurial provision models emerge, persist and co-exist and that these should be supported. This resonates with calls for international humanitarian actors to seek partnership and participation and to support local capacity.

Secondly, the paper looks at the use of the livelihoods framework by humanitarian actors and discusses concerns around its application and concepts. The superficial application of the livelihoods concept after a disaster tends to focus on households and their assets and to overlook vulnerabilities that are differentiated by access to infrastructure and access to political influence. Recent work to profile urban livelihoods offers some hope in this regard with the caveat that it takes time and can be intrusive for vulnerable groups that adopt livelihood strategies which allow them to remain hidden. Conceptually, the livelihoods framework also fails to capture the geographic and patchy qualities of vulnerability and resilience that can vary over very short distances in a city. The different ways in which households interact with urban infrastructure systems – and other critical institutions – need to be considered in an analysis of urban vulnerability, because theses modes of organization form both in response to the governance imperatives of providers and in response to the providers' perceptions of livelihoods. This produces a tendency to amplify household vulnerability for those that are or seem (and then become) least able to buffer it. The emerging discourse on urbanism and splintering cities offers some insights into the processes and structures that drive this: food for thought in the debate on humanitarian response in urban areas.

Thirdly, building on this, the paper argues that a socio-technical systems approach, which has infrastructure systems coalescing, stabilizing and reinventing themselves better to respond to change, resonates with Robert Chambers' original conceptions of household livelihood strategies that seek ways to buffer risk. A socio-technical systems approach thus serves as a useful conceptual bridge between household vulnerability and infrastructure governance because it introduces a power relationship between providers and consumers; recognizes the importance of the spatial and collective arrangements of urban infrastructure; and examines urban risk sharing beyond the governance framework's bland conceptions of state, civil society and private sector.

Finally, the paper concludes that there is a theoretical tendency for modes of humanitarian response to privilege and bypass because: urban vulnerability is difficult to understand quickly; vulnerable people and places may be hidden in ways that prejudice assessments; and there are biases in 'administrative intrusion' towards the most accessible geographically (close to existing offices), physically (easy get to in a hi-lux) and institutionally (with documentation). Added to this is the tendency for humanitarian providers to adopt modes of organizing that are motivated by risk aversion and internal convenience with a focus on distributions of packaged assets. No amount of geographic coordination across the zones of a city can make up for the fact that all actors are providing the same narrow response and, even where it is possible to target the most vulnerable with asset packages, the ultimate equity of such a response is in question because without sufficient attention to infrastructure, institutions and systems that connect the livelihoods of the most and least privileged, a transition towards recovery and resilience is unlikely.

This paper's theoretical contribution is a visual framework that allows humanitarians to ask: How are we organized, where we are now, where we are going by default and where we would actually prefer to be? Applied to humanitarian action, for example, one might ask: Do our methods of analysis push us towards measuring the individual riskiness of households; is this foreseen by our mandates; do we have systems in place to avoid the processes of privilege and bypass that give patchiness an endogenous life of its own; how can we make sure that the livelihoods of the powerful do not damage but enhance the livelihoods of others? Returning to the idea that international humanitarian actors are caught between gathering detailed livelihood profiles and intruding on vulnerable urban dwellers, the paper describes the tension between building up a highly refined and atomized understanding of urban vulnerability as a basis for apportioning costs or distributing assets and the use of more blurred and prejudiced assumptions about the riskiness of whole groups that allow costs to be unwittingly shared, sometimes in a virtuous cycle of privilege and sometimes in a downward spiral of bypass. Understanding, individualizing and quantifying risk, if that is what we should do, is only the first part of the picture. Persuading groups to share risk and give up information about themselves in a climate of splintering urbanism, individual responsibility and contingency is the challenge of humanitarian governance. Risk can only be shared when it is willingly and knowingly understood across institutional actors, across splinters of infrastructure and across livelihoods. This means building a better understanding of vulnerability in collaboration with the people at risk, such that it will not allow powerful systems to further discriminate. Faced with a changing climate, fluctuating energy and food prices and a volatile economy that will reorder the geography and nature of risk, these are critical questions for mitigating vulnerability.

References

Albert, R., Jeong, H. and Barabasi, A.-L., 2000. Error and attack tolerance of complex networks. *Nature*, 406(July 27). 378–381.

Allen, A., Davila, J. and Hofmann, P., 2006. The peri-urban water poor: citizens or consumers? *Environment and Urbanization*, 18(2). 333–351.

Batley, R., 1997. Social agency versus global determinism in Latin American urban development. *Third World Planning Review*, 19(4). 338–341.

Bhattacharjee, A., Jacob, P. M., Sumasundaram, M., Ramachandran, R., Thileepan, S., Kumar, S., Srikantharajah, S. and Suppiah, L., 2007. *Final Evaluation of CARE Australia Supported Tsunami Response in Trincomalee and Batticaloa Districts of Sri Lanka*. CARE Australia and CARE Sri Lanka, May.

Borja, J. and Castells, M., 1997. *Local and Global: The Management of Cities in the Information Age*. United Nations Centre for Human Settlemenst (Habitat) Earthscan Publications Ltd, London.

Budds, J. and McGranahan, G., 2003. Are the debates on water privatization missing the point? Experiences from Africa, Asia and Latin America. *Environment and Urbanization*, 15(2). 87–113.

Bustamente, R., Butterworth, J. and Faysse, N., 2004. *Is there a Future for Locally-managed Domestic Water Supply Systems in Peri-urban Cochabamba, Bolivia? Analysis of Performance and Some Scenarios*. Centro AGUA, UMSS. Working Paper for NEGOWAT project workshop, São Paulo, 16–21 August 2004.

Chambers, R., 1995. Poverty and livelihoods: whose reality counts?. *Environment and Urbanization*, 7(1). 173–204.

Chambers, R., 1997. *Whose Reality Counts?* Intermediate Technology Publications, London.

Chambers, R., 2005. *Ideas for Development*. Earthscan, London.

Clermont, C., Sanderson, D., Sharmu, A. and Spraos, H., 2011. *Urban Disasters – Lessons from Haiti: Study of Member Agencies' Responses to the Earthquake in Port au Prince, Haiti*. January 2010, Report for the Disasters Emergency Committee (DEC). DEC.

Crawford, C.A., 2009. *Sustainable Livelihoods and Infrastructure: Governing and Configuring Urban Water and Sanitation for Reduced Vulnerability in Cusco, Peru*. Thesis (Ph.D.), University of London, London.

Crawford, C.A. and Bell, S., 2011. Analysing the relationship between urban livelihoods and water infrastructure in three settlements in Cusco, Peru. *Urban Studies* 0042098011408140, first published on June 6, 2011. doi:10.1177/0042098011408140.

Estache, A., Gomez-Lobo, A. and Leipziger, D., 2001. Utilities privatization and the poor: Lessons and evidence from Latin America. *World Bank Policy Research Working Paper No. 2262*, 29(7). 1179–1198.

Fagen, P. W., Fernandez Juan, A., Stepputat, F. and Lopez, R. V., 2003. *Internal Displacement in Colombia: National and International Responses*. IIS/GI. Kongevej Working Paper 03.6. Institute for International

Studies, Copenhagen. www.diis.dk/graphics/CDR_Publications/cdr_publications/working_papers/wp-03-6.pdf.

Farrington, J., Ramasut, T. and Walker, J., 2002. *Sustainable Livelihoods Approaches in Urban Areas: General Lessons, with Illustrations from Indian Cases*. Overseas Development Institute, London.

Graham, S., 2000. Introduction: cities and infrastructure networks. *International Journal of Urban and Regional Research*, 24(1): 114–119.

de Haan, A., Drinkwater. 2002. *Methods for Understanding Urban Poverty and Livelihoods*. UK Department for International Development's Infrastructure and Urban Development Department.

Haiti Shelter Cluster. 2010. *Haiti: Transitional Shelter Needs Analysis*. Shelter Cluster, September 22. www.sheltercluster.org/communities/Inf.Manager/Practical%20Excercises/Transitional%20Shelter%20Needs%20Analysis.doc.

Hardoy, A. and Schusterman, R., 2000. Privatisation of water supply and sanitation services, and the urban poor. *Environment and Urbanization*, 12(2). 63–75.

Hardoy, J., Mitlin, D. and Satterthwaite, D., 2001. *Sustainable Development and Cities*. Chapter 8. Earthscan, London.

HERR, (eds.). 2011. *Humanitarian Emergency Response Review*. Independent Review Panel Chaired by Lord (Paddy) Ashdown, March. www.dfid.gov.uk/Get-Involved/Disasters-and-emergencies/Humanitarian-Emergency-Response-Review/.

IASC. 2009. *Meeting Humanitarian Challenges in Urban Areas (MHCUA) DRAFT ASSESSMENT 02.10.2009*. Inter-Agency Standing Committee, Geneva. www.humanitarianinfo.org/iasc/pageloader.aspx?page=content-subsidi-common-default&sb=74.

IFRC. 1999. *Vulnerability and Capacity Assessment: An International Federation Guide*. International Federation of Red Cross and Red Crescent Societies, Geneva. www.sheltercentre.org/library/Vulnerability+And+Capacity+Assessment+An+International+Federation+Guide.

Jacobsen, K. 2008. *Internal Displacement to Urban Areas – the Tufts-IDMC Profiling Study*. http://wikis.uit.tufts.edu/confluence/display/FIC/Internal+Displacement+to+Urban+Areas-the+ Tufts-IDMC+Profiling+Study.

Kaika, M. and Swyngedouw, E., 2000. Fetishizing the modern city: the phantasmagoria of urban technological networks. *International Journal of Urban and Regional Research*, 24(1). 120–138.

Kumar, S. 2001. *Social Relations Rental Housing Markets and the Poor in Urban India: Final report to the Infrastructure & Urban Development Department, Department for International Development (DFID)*. Printed at the London School of Economics and Political Science. http://siteresources.worldbank.org/INTURBANPOVERTY/Resources/social-relation-kumar.pdf.

Lacabana, M., 2008. *An Overview of the Water Supply and Sanitation System at Metropolitan and Peri-urban Level: the Case of Caracas*. Development Planning Unit, London.

Leipziger, D., 2003. *Achieving the Millennium Goals. The Role of Infrastructure*. World Bank, Washington, DC.

Marvin, S. and Graham, S., 2001. *Splintering Urbanism: Networked Infrastructures, Technological Mobilities and the Urban Condition*. Routledge, London.

McGranahan, G., Jacob, S. and Marianne, K., 1999. *Sustainability, Poverty and Urban Environmental Transitions*. Earthscan, London.

Misago, J. P., Guindrey, V., Duponchel, M. and Landau, L., 2009. DRAFT: University of the WitwatersrandForced Migration Studies Programme Discussion Document: Alexandra and Central Johannesburg Vulnerability Pilot Survey. A collaborative project of: The United Nations Organisation for the Coordination of Humanitarian Affairs (UNOCHA); South African Red Cross Society (SARCS); Forced Migration Studies Programme (FMSP). www.migration.org.za.

Montgomery, M. R., Reed, H. E., Satterthwaite, D., White, M., Cohen, M., McGee, T. and Yeung, Y., 2004. *Why Location Matters*. Chapter 2. Earthscan, London.

Newman, P., 2006. The environmental impact of cities. *Environment and Urbanization*, 8(2). 275–295.

Nickson, A., 1997. The public-private mix in urban water supply. *International Review of Administrative Sciences*, 63(2). 165–186.

Norwegian Refugee Council. 2008. *Camp Management Toolkit*. Norwegian Refugee Council. www.sheltercentre.org/ library/Camp+Management+Toolkit.

OXFAM. 2005. *Transitional Settlement: Displaced Populations*. Oxfam, Cambridge. www.sheltercentre.org/library/ transitional+settlement+displaced+populations.

Pantuliano, S., 2009a. Charting the way: integrating land issues in humanitarian action. *Uncharted Territory: Land, Conflict and Humanitarian Action*, S. Pantuliano (ed.). Chapter 10. Practical Action, London.

Pantuliano, S., 2009b. Going home: Land, return and reintegration in Southern Sudan and the three areas. *Uncharted Territory: Land, Conflict and Humanitarian Action*, S. Pantuliano (ed.). Chapter 8, Practical Action, London.

Payne, G., Durand-Lasserve, A. and Rakodi, C., 2009. The limits of land titling and home ownership. *Environment and Urbanization*, 21(2). 443–462.

Rakodi, C., 2002a. *A Livelihoods Approach – Conceptual Issues and Definitions*. Chapter 1. Earthscan Publications Ltd, London.

Rakodi, C., 2002b. *Conclusions*. Chapter 19. Earthscan Publications Ltd, London.

Ray, D., 1998. *Development Economics*. Princeton University Press, Princeton, NJ.

Sassen, S., 2001. *The Global City: New York, London, Tokyo*. Princeton University Press, Princeton, NJ; Oxford.

Setchell, C. A., 2005. *The Economic Impacts of Humanitarian Shelter Assistance*. USAID, Washington, DC. http://pdf.usaid.gov/ pdf_docs/PNADN547.pdf.

Sheppard, S. and Hill, R., 2005. *The Economic Impact of Shelter Assistance in Post-Disaster Settings*. CHF and USAID, Maryland, USA. www.chfhq.org/files/2136_file_EIES_final.pdf.

Solo, T. M., 2003. *Proveedores independientes de agua potable en América Latina _El otro sector privado en los servicios de abastecimiento de agua_*. The World Bank, Washington, DC.

Sphere Project. 2004. *Humanitarian Charter and Minimum Standards in Disaster Response*. The Sphere Project. www.sheltercentre.org/library/Humanitarian+Charter+and+Minimum+Standards+Disaster+ Response.

Sphere Project. 2011. *Humanitarian Charter and Minimum Standards in Disaster Response*. The Sphere Project. www.sheltercentre.org/library/Humanitarian+Charter+and+Minimum+Standards+Disaster+ Response.

Suvatne, M., 2010. Urban shelter guidelines: assistance in urban areas to populations affected by humanitarian crises. Geneva.

Swyngedouw, E. and Heynen, N. C., 2003. 'Urban political ecology, justice and the politics of scale. *Antipode*, 34(4). 898–918.

Tsunami Evaluation Coalition. 2006. *Links between Relief, Rehabilitation and Development in the Tsunami Response*. TEC, London. www.alnap.org/pool/files/lrrd-final-report.pdf.

Tsunami Evaluation Coalition. 2007. *Synthesis Report: Expanded Summary Joint Evaluation of the International Response to the Indian Ocean Tsunami*. www.alnap.org/pool/files/889.pdf.

UN-HABITAT. 2003. *Water and Sanitation in the World's Cities*. Earthscan Publications Ltd, London.

UN-HABITAT. 2006. *State of the World's Cities 2006/7: The Millennium Development Goals and Urban Sustainability: 30 Years of Shaping the Habitat Agenda*. Earthscan Publications Ltd, London. www.unhabitat.org/pmss/listItemDetails.aspx?publicationID=2101.

Unruh, J. D., 2009. Humanitarian approaches to conflict and post-conflict legal pluralism in land tenure. *Uncharted Territory: Land, Conflict and Humanitarian Action*, S. Pantuliano (ed.). Chapter 3. Practical Action, London.

van Vliet, B., Chappells, H. and Shove, E., 2005. *Infrastructures of Consumption: Environmental Innovation in the Utility Industries*. Earthscan, London.

de Waal, A. 2009. Chapter 1 Why humanitarian organizations need to tackle land issues. *Uncharted Territory: Land, Conflict and Humanitarian Action*, S. Pantuliano (ed.). Practical Action, London.

World Bank. 1994. *World Development Report 1994: Infrastructure for Development*. Oxford University Press, Washington, DC.

World Bank. 2004. *World development Report 2004: Making Services Work for Poor People*. Oxford University Press, Washington, DC.

World Bank. 2006. *World Development Report*. Oxford University Press, Washington, DC. www.rrojasdatabank.org/ wdr06final.htm.

World Bank. 2010. *Natural Hazards, UnNatural Disasters: The Economics of Effective Prevention*. The World Bank, Washington, DC. www.gfdrr.org/gfdrr/sites/gfdrr.org/files/nhud/ files/NHUD-Report_Full.pdf.

Wratten, E., 1995. Conceptualising Urban Poverty, Washington DC. *Environment and Urbanization*, 7(1). 11–36.

From research to practice (and vice versa) for post-disaster settlement and shelter

ILAN KELMAN[1], JOSEPH ASHMORE[2], ESTEBAN LEON[3] AND SANDRA D'URZO[4]

[1]CICERO (Center for International Climate and Environmental Research – Oslo), Oslo, Norway
[2]Independent Humanitarian Advisor, Brighton, UK
[3]UN-HABITAT, Geneva, Switzerland
[4]IFRC (International Federation of Red Cross and Red Crescent Societies), Geneva, Switzerland

Shelter Projects (2008 – IFRC/UN-Habitat/UNHCR and 2009 – IFRC/UN-Habitat) are two books aiming to consolidate some of the lessons from post-disaster settlement and shelter case studies, spanning over 60 projects from 1945 to the present. This paper analyses many of these case studies in order to reflect on how research and practice could be better connected for post-disaster settlement and shelter. The analysis reveals that challenges exist in compiling comparative and credible data, especially due to research needs diverging from operational needs. Methodological reflections on post-disaster settlement and shelter research are provided, using occupancy rates as a key indicator and example of the challenges faced. The main lessons provided relate to timelines. Emergency shelter programmes usually take many months, or in some cases years, to implement fully, meaning that their time frame tends to exceed that of media and donor interest. Meanwhile, research needs more longitudinal studies to provide long-term policy and practice lessons. Overall, providing disaster-affected populations with support so that they can identify and meet their own post-disaster settlement and shelter needs is the main lesson that emerges from both research and operational experience.

1. Introduction and background

Turner (1972, p. 148) identified 'housing as a verb'. It denotes that housing is often seen to be merely a physical structure in which people live. Instead, housing should be viewed as continuing actions in which the physical structure provides a focus for living in a comfortable, secure and dignified manner.

Turner's (1972) work did not apply specifically to post-disaster situations. Davis (1978, p. 33) explained, with respect to disaster contexts, that 'shelter must be considered as a process, not as an object'. When shelter is provided to a disaster-affected population, it should mean more than a structure in which people live. The shelter structure must be erected and finished in such a way that the community, the occupants' livelihoods and the wider environment are supported. Kennedy et al. (2008) summarizes the specific needs as

- Physical and psychological health including protection from the elements and a feeling of home and community.
- Privacy and dignity for families and for the community.
- Physical and psychological security.
- Livelihood support.

As knowledge and experience built up regarding post-disaster settlement and shelter, the lessons evolved into articulating and implementing the process of moving from emergency shelter to permanent housing (e.g. Shelterproject, 2003; Corsellis and Vitale, 2005). The aim of emergency shelter is to avoid life-threatening situations by meeting immediate needs. Emergency shelters, however, are often never replaced by external organizations, and so families are forced to use them for months or years or to use their own resources for permanent housing (e.g. Leon et al., 2009). Reconstruction and permanent housing, in the relatively few cases where the latter was built, should both aim to reduce vulnerability to disasters in the future.

In dealing with post-disaster settlement and shelter, the social and environmental context must be taken into account (e.g. Cuny, 1983; Quarantelli, 1995; Babister and Kelman, 2002; Ashmore et al., 2003; Barakat, 2003). Different cultures have different livelihoods, privacy standards, health needs and psychological responses to settlement and shelter processes. Meanwhile, climates vary as do other hazards along with environmental characteristics such as slopes, drainage, insects and soils, and so the settlement and shelter must be designed and built appropriately for those circumstances.

Understanding and factoring in all these contexts is not straightforward in operations, especially during or immediately after a disaster when there is intense pressure to save lives immediately, especially from other sources such as the media and donors. Consequently, minimum standards (e.g. SPHERE Project, 2004), accountability procedures (e.g. Resster, 1978; Saunders, 2004) and field guidelines (e.g. Davis and Lambert, 2002; Shelter Centre, 2010) have been published to train personnel before working on post-disaster settlement and shelter and to support them in the field.

In parallel, researchers and practitioners publish case studies in academic settings to identify lessons within the context of academic literature. Much of this work covers specific cases or specific aspects across case studies. For example, Ashmore et al. (2003) focus on post-Taleban Afghanistan, whereas Oliver-Smith (1990) uses the 1970 earthquake and avalanche in Peru. Sometimes, case studies are compared such as Kennedy et al. (2008) looking at Sri Lanka and Aceh after the 2004 tsunami. Johnson et al. (2006) compare Turkey and Colombia after earthquakes in 1999.

Large-scale emergencies such as Hurricane Mitch in Central America in 1998 or the 2004 Indian Ocean tsunami often have operational responses documented by research institutes such as Groupe URD (e.g. URD, 1999, 2005). Yet, agencies rarely return to places in which they have operated to review the impacts of their post-disaster settlement and shelter activities over time, especially to examine how their work supported, or did not support, long-term recovery.

Consequently, most studies and reports are one-off, rather than examining the same location over time. Snarr and Brown (1980, 1982, 1994) represent one of the few longitudinal case studies of post-disaster settlement and shelter. They examined occupant expectations and satisfaction over time regarding the housing received after Hurricane Fifi struck Honduras in 1974.

In terms of specific aspects of post-disaster settlement and shelter, Karunasena and Rameezdeen (2010) look at data to compare the strengths and limitations of owner-driven and donor-driven post-disaster shelter programmes. Using four case studies of post-disaster housing, Davidson et al. (2007) concentrate on the strengths and limitations of community participation. Leon et al. (2009) focus on using post-disaster settlement and shelter for capacity building, both within the sector internationally and for disaster-affected people.

All these studies are needed and are legitimate, forming a research foundation on which lessons for policy and practice do and should emerge. For instance, by studying only the economic impact of post-disaster shelter, Sheppard and Hill (2005) demonstrate the importance of post-disaster settlement and shelter for creating and maintaining sustainable livelihoods. Yet, the literature provides few meta-research papers that aim to cover a wide range of topics across a wide range of case study circumstances and years. Meanwhile, aid agencies rarely analyse the impact of their post-disaster settlement and shelter projects.

A few examples of literature and agency reports that cover such work are Davis (1977, 1978), Hayles (2010), Shelter Centre (2010) and UNDRO (1982), which consolidate case studies and experience into principles and practices for practitioners in the field. Similarly, Setchell (2006) summarizes the main lesson from many post-disaster shelter case studies of needing to think about the long-term when implementing any project.

Within the context of this work, Shelter Projects (2008, 2009) are two books, aiming to consolidate some of the lessons from post-disaster settlement and shelter case studies. These case studies are the outcome of an institutional collaboration among three agencies with specific roles in providing post-disaster settlement and shelter: the International Federation of Red Cross and Red Crescent Societies (IFRC), UN-Habitat and United Nations High Commissioner for Refugees (UNHCR). Most of the case studies and the data from them have not been published before, making these books a valuable reference of practical experience with a basis in the academic literature, as referenced above.

To build on this work, this paper adds to the few publications consolidating findings from many post-disaster settlement and shelter case studies. It takes a broad-brush approach through exploring numerous case studies in order to better connect research and practice. Rather than being a comprehensive literature review or a formal review paper, the purpose of this paper is to provide an overview of several post-disaster settlement and shelter case studies from past decades in order to reflect on how research and practice could be better connected. This paper is taken as an opportunity to analyse case studies presented in Shelter Projects (2008, 2009), since those books did not have the opportunity to do a comparative analysis, to critique the field, or to extract more general lessons learned.

In order to allow the limitations and difficulties of case studies to be openly shared without directly criticizing individuals and organizations who did the best that they could in the field with limited resources and under difficult circumstances, none of the case studies here (or, for example, in Shelter Projects, 2009) are directly attributed to specific organizations. As a result of the projects being implemented in diverse and often challenging conditions, they illustrate both good and bad practices. From every case study, lessons are available to be learned, yet not all are transferable to all contexts, thus recognizing which aspects should be repeated or avoided elsewhere is needed.

2. Case studies

2.1. Overview of case studies

Tables 1 and 2 list the case studies from Shelter Projects (2009) used in this paper and, for Table 1, details on the project. These case studies make Shelter Projects (2009) an important training and reference resource for humanitarian workers and researchers covering post-disaster settlement and shelter. Examples of cases that do have some literature are Aceh in 2004 (e.g. Kennedy et al., 2008), Kenya in 2008 (Mwiandi, 2008) and Peru in 1970 (Oliver-Smith, 1986, 1990). Examples of case studies in the literature that are not included here are Mozambique after the 2000 floods (Gall, 2004), Turkey after the 1999 earthquakes (e.g. Johnson, 2007) and the USA after Hurricane Katrina in 2005 (Nigg et al., 2006).

Case studies included here were all implemented on a large scale in the field. Trials or design concepts are not mentioned here. The case studies were selected using the following criteria as a guide:

- The post-disaster settlement and shelter project was implemented in full.
- A minimum of 500 families had improved shelter as a result of the project.
- The project was implemented largely within the first year following non-conflict disasters. For conflict-affected populations, chronic emergencies, returns processes and longer timescales were considered.

TABLE 1 List and categories of case studies

Country	Year	Type of emergency	Urban?	Non-food items		Shelters built		Cash voucher	Labour			Technical capacity
				Household items	Shelter materials	Post-emergency	Permanent		Community	Contracted	Direct	
Afghanistan	2002	Conflict/ complex		✓	✓		✓	✓			✓	✓
Afghanistan	2009	Conflict/ complex			✓	✓			✓			
Azerbaijan	1992	Conflict/ complex	✓			✓				✓		
Bangladesh	2007	Cyclone		✓	✓	✓		✓	✓	✓	✓	✓
Bangladesh	2009	Cyclone		✓		✓					✓	✓
China, Sichuan	2008	Earthquake			✓		✓	✓	✓	✓		✓
DRC, Goma	2002	Volcano	✓	✓		✓		✓	✓		✓	✓
DRC, Goma	2009	Conflict/ complex		✓	✓		✓	✓	✓			✓
Eritrea	1998	Conflict/ complex		✓	✓							
Gaza, Palestine	2009	Conflict/ complex	✓									✓
Georgia	2008	Conflict/ complex		✓	✓		✓	✓			✓	
Haiti	2008	Flooding	✓	✓	✓	✓						✓
India, Gujarat	2001	Earthquake			✓			✓	✓		✓	✓
Indonesia, Aceh	2004	Earthquake, tsunami	✓				✓		✓	✓		✓
Indonesia, Jogyakarta #1	2006	Earthquake						✓				
Indonesia, Jogyakarta #2	2006	Earthquake		✓		✓			✓		✓	✓
Italy	2009	Earthquake	✓			✓				✓		
Kenya	2007	Conflict/ complex			✓		✓		✓		✓	✓
Kenya	2008	Conflict/ complex			✓	✓						✓

Continued

TABLE 1 Continued

Country	Year	Type of emergency	Urban?	Non-food items		Shelters built		Cash voucher	Labour			Technical capacity
				Household items	Shelter materials	Post-emergency	Permanent		Community	Contracted	Direct	
Kenya	2009	Conflict/ complex			✓		✓		✓		✓	✓
Liberia	2007	Conflict/ complex			✓		✓	✓	✓			✓
Mozambique	2007	Cyclone			✓			✓				
Myanmar	2008	Cyclone			✓		✓	✓			✓	
Pakistan #1	2005	Earthquake		✓	✓			✓	✓		✓	
Pakistan #2	2005	Earthquake				✓			✓			✓
Peru #1	2007	Earthquake	✓		✓	✓			✓			✓
Peru #2	2007	Earthquake	✓			✓			✓		✓	✓
Peru #3	2007	Earthquake				✓				✓		✓
Russia, Ingushetia	1999	Conflict/ complex	✓			✓		✓	✓	✓	✓	
Rwanda	2008	Conflict returnees			✓		✓				✓	✓
Somalia	2007	Conflict/ complex	✓				✓			✓		✓
Somalia	2008	Conflict/ complex	✓	✓	✓	✓					✓	✓
Somalia #1	2009	Conflict/ complex	✓		✓	✓	✓				✓	✓
Somalia #2	2009	Conflict/ complex	✓		✓	✓					✓	✓
Sri Lanka	2004	Tsunami				✓					✓	
Sri Lanka	2007	Conflict returnees				✓	✓	✓	✓			✓
Sudan	2004	Conflict		✓	✓							
Uganda	2007	Slow-onset floods			✓				✓			✓

Complex refers to factors in addition to the conflict, such as drought and returnees from other locations, in addition to the sudden onset shelter need resulting from conflict.

TABLE 2 Historical case studies

Country	Year	Type of emergency
Algeria	1980	Earthquake
Bangladesh	1975	Conflict
Guatemala	1976	Earthquake
Haiti	1982	Analysing a document on the status of shelter
Honduras	1974	Hurricane
Honduras	1998	Hurricane
India	1971	Conflict
India #1	1977	Cyclone
India #2	1977	Cyclone
Italy	1976	Earthquake
Nicaragua #1	1972	Earthquake
Nicaragua #2	1972	Earthquake
Peru	1970	Earthquake
Sudan	1985	Conflict
Thailand	1979	Conflict
Tonga	1982	Cyclone
Turkey, Caldiran	1976	Earthquake
Turkey, Gediz	1970	Earthquake
Turkey, Lice	1975	Earthquake
UK	1945	Post-conflict
Yugoslavia (former)	1963	Earthquake

These case studies had limited information available, provided in Shelter Projects (2008, 2009), but assisted in putting contemporary case studies into wider contexts.

Also, enough project information had to be available to create a case study, either from staff involved or from original documentation available. In many cases, information was incomplete. For the case studies in Table 2, representing more historical projects, information tended to be less documented than for the more recent case studies in Table 1. For the case studies in Table 1, the information was built up by contacting personnel who were involved in the project and by corroborating their information with written documentation that they could provide or that could be found elsewhere.

Furthermore, the selected case studies deliberately illustrate a diversity of approaches to meet post-disaster settlement and shelter needs. The range of projects includes building damage assessment (Gaza in 2008), cash or voucher distribution (Goma in 2009), hiring contractors to build shelters (Azerbaijan in 1992), supporting families in collective centres to find rental properties (Haiti in 2008) and training for improving one's settlement and shelter (Uganda in 2007). Many of the projects, such as Bangladesh in 2007, used a combination of approaches to meet identified needs as they evolved after the disaster.

In all the cases, providing shelter is more than simply designing and building structures. Agencies are moving away from the relief 'truck-and-chuck' approach, in which shelter items such as tents or tarpaulins are brought in and distributed without proper understanding of the contextual needs. Instead, agencies are trying to implement more tailored and more appropriate assistance.

Therefore, as a result of the context and needs, nothing was directly constructed in some of the projects. An example is Goma in 2009 where traditional hosting systems were supported. In other projects, the advocacy and learning support components of the project had a larger impact than the organization was able to achieve through construction alone. Bangladesh in 2007 included training on safe shelter awareness and repair along with technical support for repair and maintenance to help families learn how to keep their shelter in good condition. Aceh in 2004 is an example of advocacy being used to support land rights, especially given the post-conflict situation and the number of land rights documents destroyed in the tsunami. The projects also illustrate the importance of adaptability, such as Georgia in 2009 demonstrating how projects must be continually monitored and adjusted to meet the changing context.

In almost all case studies, the first and main effort regarding post-disaster settlement and shelter came from the affected people themselves. Of the case studies, the more effective projects were implemented with the close involvement of the affected people. In Peru in 2007, one local NGO used neighbourhood public kitchens to mobilize communities for distributing re-usable construction materials for those most in need.

Technical support was provided in the form of a manual that had been written before the earthquake, along with carpenters who helped to train the affected population. The involvement of the population was facilitated by the implementing NGO having been on the ground in the communities before the disaster, so it had the population's trust and knew the communities.

In some case studies, institutional frameworks and the role of governments can become prominent. In Italy after the 2009 earthquake, although there was strong cooperation among local authorities, contractors and the disaster-affected population, most of the decisions were government-led within a prescriptive legal framework that limited inputs from the affected population. Similarly, for China in 2008, the earthquake damaged an estimated five million dwellings, but a strong central government lead, bolstered by the country's centralized political system, facilitated the post-disaster settlement and shelter.

Figure 1 illustrates the duration of the case studies in Table 1 along with the time delay from the height of the disaster until when the post-disaster settlement and shelter project began. Some projects started during emergency response, such as supporting Chechen displaced people in Ingushetia in 1999 as the conflict continued. Other projects were not implemented until several months after emergency response had been completed. Implementation in Gaza in 2009 started 3.5 months after the war stopped.

Some case studies were a response to an ongoing chronic disaster, such as long-term conflict, drought or a combination. As such, calculating duration until project start is not possible. Other case studies are not included in Figure 1 because they were a continuation or update of previous projects. Kenya in 2009 was a separate operational project with a different budget from Kenya in 2008, yet the 2009 project continued the work done for the 2008 project. Eritrea in 1998 and Afghanistan in 2002 are not included in Figure 1 because the responses summarized were part of ongoing emergency programmes. In the case of Afghanistan in 2002, elements of the same programme continue today.

Some of the projects took years to complete, especially those that were implemented on a large scale such as Aceh in 2004 for which work was still ongoing 3 years after the disaster despite a lasting post-tsunami peace. The project in Rwanda in 2008 took 2 years to build 220 houses. In most cases, permanent housing construction takes many years, as complicated issues such as land ownership need to be resolved before construction can even begin. This leaves open the question of where families should live as emergency shelters decay. Into this gap, organizations often become involved in 'transitional settlement and shelter' construction, building simple structures that will provide improved shelter until more permanent solutions become available (e.g. Corsellis and Vitale, 2005; Leon et al., 2009; Rand et al., 2011).

The end state of the 'transition' is not always clear. The example epitomizing this situation is post-war UK in 1945: many of the 156,000 houses that were built as temporary accommodation for returning soldiers and their families in 1945 are still occupied today. Consequently, and wherever possible, post-disaster responses should start considering immediately how to support permanent housing and incremental approaches, even while setting up emergency options. Bangladesh in 2007 shows to some degree how incremental approaches towards permanency can be cost-effective for agencies while generating income and a sense of ownership among the disaster-affected population, as people are given the opportunity to improve and maintain their own shelters.

Commonly, following a large-scale disaster, a surge of media attention occurs at the outset, with an initial focus on search and rescue and then on aid delivery. The level of media interest often provides impetus for the pace of relief distributions, but media interest can rapidly fall off even as aid delivery continues or as more assistance is needed – especially assistance in moving from emergency to permanent settlements and shelter. Figure 2 illustrates the speeds of

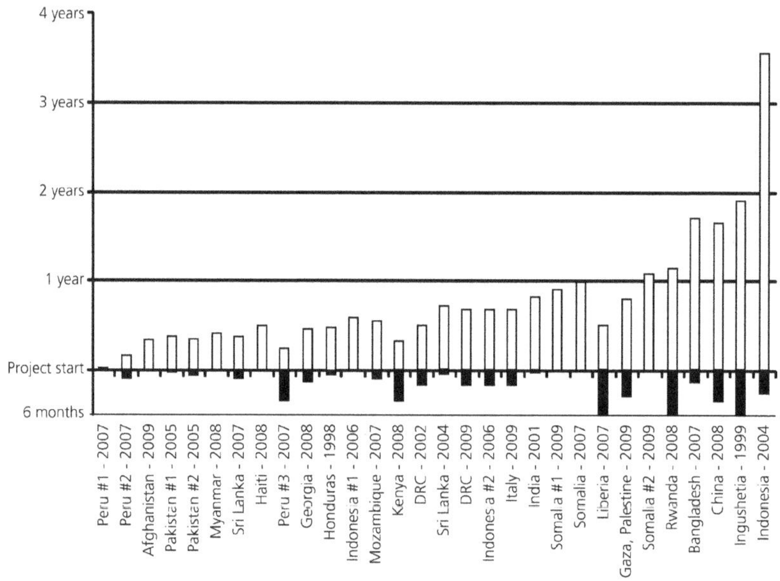

FIGURE 1 Duration of shelter projects (from Table 1), and the time between the height of each disaster and the start of the shelter project (courtesy of IFRC and UN-Habitat, from Shelter Projects, 2009).

emergency aid from responses for five projects across four disasters. The timeline should be considered in comparison to the duration of media interest usually being days or weeks.

Relief distributions are inevitably run on longer time frames than the initial media reporting of the emergency. In the first month, organizations must often rapidly scale up staffing, establish supply pipelines and mobilize distribution teams along with agreements with and support from partner, particularly local, organizations. From the examples in Table 1, one of the quicker distributions of shelter items was Jogyakarta in 2006, which lasted almost 4 months. That time frame is long beyond the ability of a disaster to sustain international headlines.

Post-disaster shelter programming usually takes even longer to implement than the immediate distribution of emergency supplies (Figure 2). The post-disaster shelter response for Jogyakarta in 2006 took 12 months to build 75,000 shelters as a basis for more permanent housing. That is still relatively quick and was achieved only because local, lightweight shelter materials were used, such as bamboo. That speed would not have been possible through using imported, bulky materials. Using local materials also raised concerns about local deforestation and resource overuse, as discussed for Aceh in 2004 by Greenomics Indonesia and WWF Indonesia (2005).

The literature overview in the previous section noted that post-disaster settlement and shelter needs to be environmentally contextual. The case studies presented here demonstrate how the projects responded to different hazards. In the town of Bosaso in Somalia in 2009, one of the major threats to families was fire, which had destroyed many people's homes. The best way to reduce the risk of a major fire in the temporary settlement was to work with communities to improve solid waste disposal, to install fire breaks and to establish fire prevention committees. This proved more cost effective and pragmatic than trying to build fire-proof shelters. No major fire occurred during the time of the project.

Sozma Qala camp in Afghanistan in 2009 illustrates climate hazards. Tents and water supplies were winterized prior to the onset of winter.

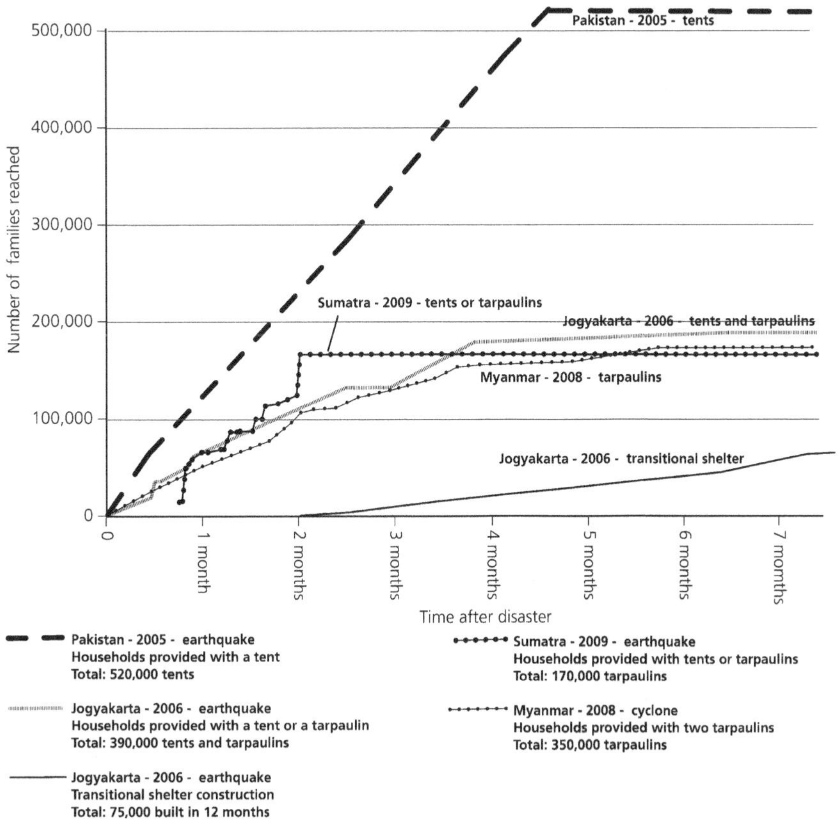

FIGURE 2 Graph of shelter materials distributed to households against time for four major disasters[1] (courtesy of IFRC and UNHabitat, from Shelter Projects, 2009).

Following floods in Uganda in 2007, the post-disaster shelters included flood risk reduction measures along with measures to maintain thermal comfort indoors. For example, the shelter and foundations were elevated and a waterproof barrier was put in the foundation to protect the walls and floors, which were made of adobe blocks to maintain indoor air temperature. Termites were also a major problem, and so wood in direct contact with the ground was treated to protect it from termites. The process of building the shelters not only provided the occupants with less vulnerable dwellings, but also trained the occupants in maintaining these measures and in explaining why the flood and termite risk reduction techniques were implemented.

2.2. Constraints

Figure 3 shows the range of shelter areas covered by different shelter projects. China in 2008, Myanmar in 2008, Palestine in 2009, Somalia in 2008 and Somalia in 2009 #1 are not included because those projects did not build shelters directly, and so size cannot be calculated. The disparity among projects in Figure 3 results from varying needs, level of permanency of the shelter, budget and logistics constraints, host standards and official policies.

Despite these many factors, funding tends to be the main determinant of the types of post-disaster settlement and shelter response for emergency or permanent needs. Funding levels are

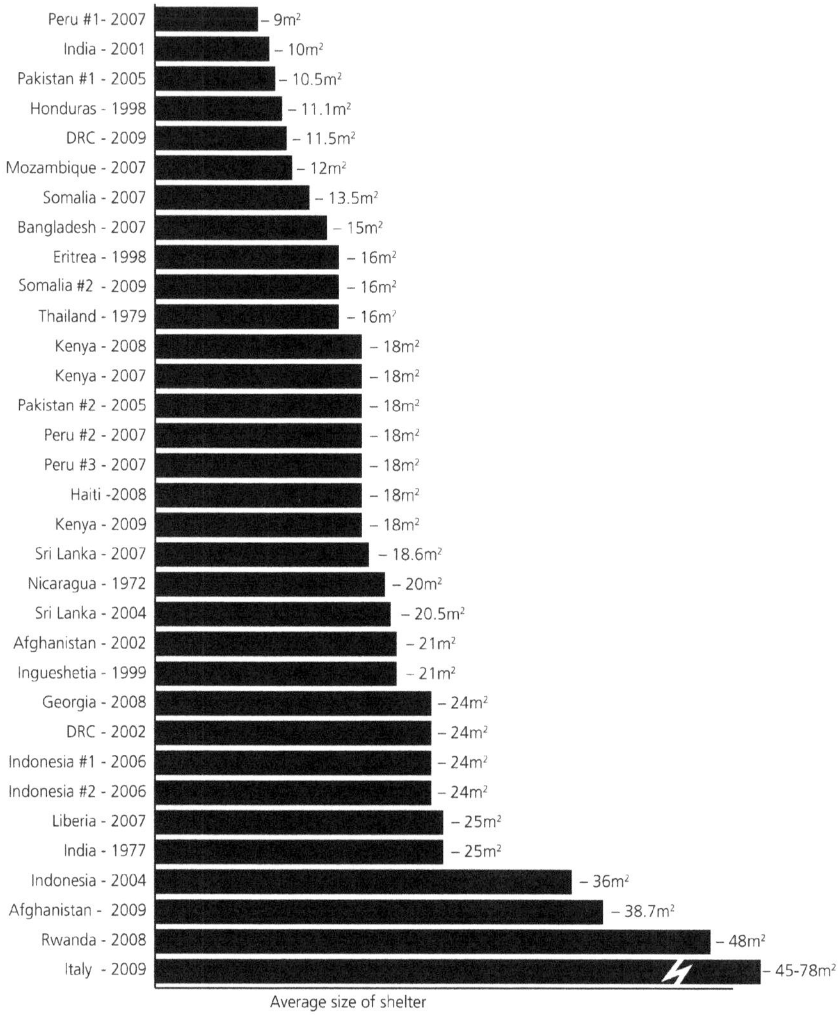

FIGURE 3 The range of shelter areas covered by different shelter projects (courtesy of IFRC and UNHabitat, from Shelter Projects, 2009).

rarely concomitant with the population's needs (e.g. Polman, 2010). A good illustration is the contrast between the responses in Italy in 2009, which provided up to 78 m^2 per family, and one of the projects in Peru in 2007, which provided 9 m^2 per family, whereas two other projects for the same disaster provided double that (Figure 3). The different responses in Bangladesh in 2007 and 2009 are partly the result of differing funding. Shelters after Cyclone Sidr in 2007 provided an average of 15 m^2 per family, whereas shelters after Cyclone Alia in 2009 consisted mainly of plastic sheeting covering small land areas on embankments – and large numbers of families were not supported by any post-disaster settlement or shelter projects at all.

The time frame of funding, not just the amounts, also impacts what can be achieved for post-disaster settlement and shelter. Emergency relief funds rarely last for more than one year, meaning that there is time pressure to complete construction of agreed numbers of shelters rather than to ensure that shelters are built appropriately, are occupied and meet the disaster-affected people's needs (even if they meet donors' needs). Funding can be earmarked, leaving more for emergency needs than for permanent approaches, even while the disaster-affected population and implementing agencies need and are ready for permanent settlement and shelter.

Given the funding constraints, a trade-off can occur between choosing adequate-quality projects for fewer people or poorer-quality responses to support more people. The case studies in Somalia illustrate this situation. In 2009, one project delivered improved but basic shelter with a limited lifetime to over 24,000 people. In 2008, a separate project had provided durable sites and services (not shelter materials), but for only 700 people over 2 years and with significantly higher project costs per person assisted.

Another major constraint is project location. Factors include where need is greatest, where organizations or their partners have existing projects, where there are gaps in settlements and shelters, avoiding hazards and where land and resources are available or are permitted to be used. Giving support to a selected location can create a pull factor and increase the population requesting support, especially if the location is similar to or better than where people were living before.

Properly assessing, establishing transparent and verifiable criteria regarding, and subsequently selecting which individuals or families will benefit from a project is one of the largest project challenges. If badly managed, it can be open to inequities, which can result in tensions and open conflict or abuse.

To reduce some such abuses, in Gonaives, Haiti in 2008, teams surveyed families in the collective centres between 3 a.m. and 4 a.m. to ensure that those surveyed were in fact resident in the shelters, rather than turning up during the day for free help. In Bangladesh in 2007, families were selected with the agreement and help of community committees. This approach ensures local knowledge, but it took several weeks before the project could start and it might also reinforce the marginalization of those needing help who were marginalized in the community before the disaster. Basing selection solely on whether a family's house is damaged can bias responses away from those who rent or squat in their homes.

3. Methodological reflections on post-disaster settlement and shelter research

3.1. How are case studies interpreted?

Operational field workers have a wealth of experience by which they can inform research. Their observations do not always conform to standard, disciplinary research methodologies. That is especially the case since research priorities in humanitarian emergencies are frequently significantly divergent from operational ones. Research can be conducted, but should not adversely interfere with the operations (Kelman, 2005), interviewees have limitations to what they should be subject to (Rosenstein, 2004), and objectivity and neutrality might not be feasible (Silkin and Hendrie, 1997). Within these contexts, this section intersects the review of case studies from the previous section with case study research methodologies to provide some reflections on methodological challenges of field research into post-disaster settlement and shelter.

When dealing with case study research, a 'unit of analysis' is often suggested as being necessary, the (somewhat circular) definition being 'the fundamental problem of defining what the "case" is' (Yin, 2003, p. 22). An implicit expectation exists that all units of analysis can be logically selected in order to conform to a neat structure with enough data available to permit unambiguous comparison. That is an ideal that should be sought, but Yin (2003) recognizes that the ideal is not always possible or needed. For example, Yin's (2003) Box 5A describes how one book 'is an excellent example of a case study' but does not give a unit of analysis 'Because the book is not an academic study, it does not need to, nor does it' (p. 25).

Yet, Yin (2003) and case study research explicitly note that the power of using case studies is to answer the questions 'why' and 'how'. Much operational research in post-disaster settlement and shelter (e.g. Ashmore et al., 2003; Manfield et al., 2004; Johnson et al., 2006; Nigg et al., 2006; Kennedy et al., 2008; Leon et al., 2009) uses case studies, effectively seeking to answer a variant of the questions 'Why is post-disaster settlement and shelter implemented the way it is seen to be implemented?' and 'How could the situation improve?'

Many of these studies do not explicitly identify a unit of analysis. As per Yin (2003), are they not academic studies despites being published in peer-reviewed journals? Do they display a significant methodological weakness? Or is the case study approach without a clear or explicit unit of analysis dictated by the necessity of post-disaster situations and studies?

The project for Eritrea in 1998 involved distributing, to temporary settlements in three locations, non-food items and other shelter items, such as tents, tarpaulins and fuel-efficient stoves. The provision of durable shelter items was not possible due to a political interest in ensuring that the settlements were temporary. As a result, the conflict-affected people often adapted the emergency shelter items they received in order to improve their living conditions. Meanwhile, to encourage return, the government of Eritrea has been providing to some returnees cash grants, two hectares of land, and assistance with building homes and buying seed and livestock.

Possible units of analyses could be the temporary settlements, households overall, households receiving each type or assistance, or returnee individuals, families or households as a case study compared to non-returnee individuals, families or households. To understand why this project selected certain items to distribute in certain locations entails an understanding of all these aspects along with the history of why the people were displaced and the politics of keeping the settlements temporary compared to the speed of return. Such characteristics exemplify why case study research is needed and Eritrea in 1998 represents a case study (e.g. Merriam, 1998; Flyvbjerg, 2006). Yet, the unit of analysis is not clear and could add confusion to the analysis if different case studies, each with its own unit(s) of analysis, are defined within 'Eritrea in 1998'.

Similarly, for Georgia in 2008, 1,850 families had their homes damaged, which were mainly single-family houses and multi-unit apartment buildings. Materials were distributed aiming for repairs that would suffice for families to stay in their houses for the winter by providing one warm room. Then, the project added a 'one warm cottage approach', building a new structure that would suffice for the winter and that families could keep.

To understand the reasons for the homes being damaged, the challenges of working on post-disaster settlement and shelter, and why programming decisions should be made, the unit of analysis is ambiguous. Should 'one warm cottage' recipients be separated from 'one warm room'? Should the unit of analysis be families from single-family houses compared to multi-unit apartment buildings? Should the unit of analysis be homes damaged but liveable compared to homes destroyed, so unliveable?

The reality from the field is that case studies and units of analysis can be defined in multiple ways. All need to be considered along with the histories of how that situation was reached. In particular, for connecting research and practice in post-disaster settlement and shelter work, the focus is often trying to put decision-making in the context of the situation faced in order to understand the factors and reasons that led to selecting particular responses. Were the responses sufficiently informed on (available) options? If decisions were dictated by limited funding, why was funding inadequate? How much organizational inertia and blinkers are involved, either from government- or donor-imposed approaches or from an agency's own remit or default operation mode? In trying to grasp these field concerns in the midst of a humanitarian emergency, it might not always be feasible to create a neat research package that conforms to methodological textbooks.

3.2. Considering occupancy rates

Operationally, humanitarian workers approach case studies in different ways each time. They act in a manner that depends on the resources available, mandates of all parties involved and the immediate field context. Most of the workers are on short-term contracts or are seconded and so tend to leave the site soon after the emergency work is completed. Visiting a case study after an international response means that most of the people who documented the response and who have the paperwork are no longer there and might not be contactable because they are working with another employer in a different location. The grey literature is full of reports with useful advice that might be hard to access, such as Lewis and Mokrane (1981) covering reconstruction of human settlements for the case study of Algeria in 1980.

Reports might or might not have been filed and retained. The reports available might or might not contain data needed to effect a comparison across many case studies. The data were often collected under intense time and resource pressures, often coupled with security concerns. Triangulating with second- or third-hand sources can be completed, but care is also needed in interpreting data and material that might not have been obtained through direct field work. This is the reality facing case studies in post-disaster settlement and shelter, as shown by Shelter Projects (2008, 2009) in the attempts to compile comparable case studies. Consequently, attempts to have consistent comparators, for evaluating projects or for comparing case studies, might not succeed to the standards of research in more controlled and controllable environments (see also Barakat and Ellis, 1996).

One simple numeric indicator of the effectiveness of post-disaster settlement and shelter is occupancy rate. These data are difficult to collect systematically because definitions of occupancy differ depending on context. For instance, if livelihoods are pursued through weekly or monthly commuting, how should the occupancy rate be calculated? Similarly, elderly relatives can be moved around with different branches of the family each taking their turn to host the elders.

As well, the numbers in each settlement and shelter change over time, not just through births and deaths which can be accounted for through re-surveying occupancy rates over time. If livelihood opportunities that are perceived to be better arise elsewhere, then many occupants will migrate away as was witnessed for Bangladesh in 2009. If plenty of resources are available, post-disaster settlements and shelter can attract migrants from other locations who were not affected by the disaster. That also means that studies in different locations or at different times in relation to the project's start might not be comparable with respect to occupancy rates; for instance, correlating occupancy rate with resident satisfaction.

An added factor is whether or not people assisted should be permitted to rent out their donated house, or some space in it, in order to generate income. If houses are rented out, then the owner might not be living there in order to be able to rate satisfaction. If rooms are rented out, then living in close proximity to strangers could reduce satisfaction balanced by extra income increasing satisfaction.

Even re-surveying occupancy rates or other data over time can be challenging. For example, it would be relevant to determine the impacts of distributing plastic sheeting or a tarpaulin to households. Field experience shows that, even several months later, distributed materials might no longer be visible or traceable. Perhaps they have decayed, been sold, been cut up or been covered up. Households might not remember what they received when. Humanitarian workers plus any records have frequently moved on. Families also move their temporary shelters to other locations over weeks and months, and so re-surveys months later might not be matchable with the original survey.

3.3. *Accepting reality in research*

The importance of context in leading to differences among possible comparators must be considered. The two case studies of the 2004 tsunami, Aceh and Sri Lanka, demonstrate this point. Numerous similarities emerged, such as the presence of a tsunami–conflict interaction and the overwhelming amount of donor support. Yet, the response was significantly different due to other factors such as national governance, local culture, who was killed by the tsunami and the conflict dynamics. Similarly, Bangladesh in 2009 and in 2007 along with all four Somalia projects displayed significant differences due to donor response (Table 1). These case studies are still comparable, but the differences must be openly discussed along with recognizing that consistency across case studies for the comparison might be an unachievable ideal.

Reality has legitimacy and coherence in its own right. Lessons should be sought by scientifically analysing the available data as best as feasible, even if that means that the analysis does not, and cannot, conform to any ideal in a textbook. That does not mean that the operational field research is entirely divorced from the textbook. Instead, considering case study research methodologies, it embraces Flyvbjerg's (2006) defence of how science can be advanced through case studies while returning to Yin's (2003) fundamental definition that a case study 'investigates a contemporary phenomenon within its real-life context' (p. 13). In a real-life context such as a humanitarian emergency, Merriam's (1998) characteristics are fully met in terms of not having control over 'a contemporary set of events' (p.32) and 'the variables are so embedded in the situation as to be impossible to identify ahead of time' (p. 32).

Many of the field studies referenced in Section 1 demonstrate the variety of disciplines and academic journals that publish this approach to research for post-disaster settlement and shelter. They demonstrate how operational work, that is, the practice, can become research, that is, through peer-reviewed papers. That represents the practice-to-research direction. The research publications then provide policy and practice recommendations with the hope that the science will lead to improvements in the field. That hope is often fulfilled, given that many authors are post-disaster settlement and shelter practitioners.

4. Conclusions: Lessons to highlight

The research–practice connections in post-disaster settlement and shelter yield several recurring lessons. The first major challenge relates to timelines of post-disaster settlement and shelter. Post-disaster shelter and settlement responses are driven mainly by external funding and not either by the needs of the disaster-affected people or by the relevance of the funding and other donations to the situation being experienced. Much is media driven which means that resources are generally available according to the media's short timeline of interest in a disaster.

That reality is unlikely to change, suggesting the need to try to take advantage of the initial 'media interested' phase for providing disaster-affected populations with settlement and shelter in the most permanent way possible within the circumstances. That can lead to a contradiction. As exemplified by Somalia in 2008 and 2009, noted above, a trade-off occurs with the number of people who can be helped compared to the quality of assistance that can be provided. Where lives are immediately at risk, time does not exist to build houses – especially with proper consultation and planning – when tents and plastic sheeting can be handed out immediately.

A further challenge arises in that research cannot provide all the answers due to large gaps in research that has previously been supported. A significant need exists for longitudinal studies, such as on resident satisfaction (Rand et al., 2011), resident adaptations to shelters (Ashmore et al., 2003), maintenance and patterns of disaster recurrence. As noted earlier, very few studies are longitudinal or have any long-term follow-up with the disaster-affected people. Getting longitudinal results takes time, but the few studies that exist (e.g. Snarr and Brown, 1980, 1982, 1984) provide a solid method to start emulating.

Another major direction to highlight for the field emerges from comparing historical (Table 2) and more contemporary (Table 1) post-disaster settlement and shelter projects. One particularly important trend to highlight over the decades is urbanization. Since 2008, over half of the world's population has been living in urban areas (UNFPA, 2007) and it appears that one long-term impact of disasters, including conflict, is to drive people further into cities. Conversely, disasters in cities tend not to drive people back into rural areas.

Post-disaster settlement and shelter responses in urban areas are becoming increasingly common, with recent examples being the 2010 Haiti earthquake; the 2004 tsunami and earthquake in Banda Aceh, Indonesia; the 2005 hurricane in New Orleans, USA; and the 2005 earthquake in Balakot and Muzaffarabad in Pakistan. More investigation into urban case studies is therefore important to consider alongside the history of predominantly rural case studies in the literature.

The main operational lesson that can be stated with confidence based on research is that post-disaster settlement and shelter operations need to be geared towards supporting people in making their own post-disaster settlement and shelter decisions. That is usually constrained by the funding available and, sometimes, by the necessity of dealing with life-threatening emergencies. That includes providing support that permits people to pursue their own flexibility and adaptability for settlements and shelter, again within some constraints, namely ensuring that the four needs identified in Section 1 from Kennedy et al. (2008) are met for everyone. For instance, that might mean avoiding discrimination against certain groups within the disaster-affected population, even if that discrimination were systemic prior to the disaster.

The challenge of creating better prepared communities, reducing individual and community vulnerability to disasters and developing more effective post-disaster assistance methods becomes increasingly poignant as more field knowledge is obtained and consolidated (e.g. in Shelter Projects, 2008, 2009). The experience exists and is being documented, yet many mistakes are repeated plus such documents do not always have the opportunity to critically analyse the material presented. Field-focused guidance documents are continuing the trend towards supporting communities for their own decision making for post-disaster settlement and shelter, while integrating those decisions with other sectors such as health, water, sanitation and education. It is an open question regarding whether or not this makes the field process too cumbersome for effective action and whether or not discrimination could be perpetuated if the community is given full decision-making power.

In the case study of Peru in 1970, locals actively sought to continue the pre-disaster social stratification through occupancy patterns of post-disaster settlements and shelter (Oliver-Smith, 1990). For Aceh in 2004 and Sri Lanka in 2004, Kennedy et al. (2008) warn that involving communities in full and fair participation for decision-making does not mean giving communities complete control of decision-making.

A balance must also be sought with the media timeline that makes resources available and that sometimes donations available are not needed. Many studies (e.g. ALNAP, 2006; Polman, 2010) point out that much humanitarian work, which includes the post-disaster settlement and shelter sector, is usually driven by the donations that are available rather than by the disaster-affected population's needs. A major research gap is describing how the media-driven, supply-driven humanitarian sector could change in order to be more responsive to the needs of the disaster-affected population while garnering sufficient resources beyond the media timeline for disasters.

It is particularly important to have the time available to work through integrated approaches to post-disaster settlement and shelter, especially when implementing permanent approaches. That means thinking beyond house-to-house construction and connecting the shelter to wider settlement needs, namely town planning, land use and infrastructure services such as water supply and waste management. These physical needs also should be linked to social and livelihood services such as law enforcement, providing opportunities for small, local enterprises, countering discrimination and education.

Most case studies in the literature do not report good practice across all these elements. The reasons are mainly lack of resources and lack of political interest in considering beyond the emergency. While research continues to improve based on practitioner experience, practitioners in the field are doing their best to apply the lessons from research but are often stymied by those with the power to change who are not learning the lessons from research.

Note

1. Figure 2: In all disasters additional materials were distributed e.g. in Pakistan corrugated iron, plastic sheeting and shelter repair kits were distributed. The information in the graph does not take into account targeting, or other support activities such as training, advocacy and voucher distribution. Data is taken from shelter cluster commodity tracking lists or from OCHA sitreps and is subject to errors in reporting.

References

ALNAP, 2006. *Key Messages from ALNAP's Review of Humanitarian Action*. ALNAP (Active Learning Network for Accountability and Performance in Humanitarian Action), London.

Ashmore, J., Babister, E., Corsellis, T., Fowler, J., Kelman, I., Manfield, P., Mcrobie, A., Spence, R., Vitale, A., Battilana, R. and Crawford, K., 2003. Diversity and adaptation of shelters in transitional settlements for IDPs in Afghanistan. *Disasters*, 27(4). 273–287.

Babister, E. and Kelman, I., 2002. The Emergency Shelter Process with Application to Case Studies in Macedonia and Afghanistan. *Journal of Humanitarian Assistance*, January. www.jha.ac/articles/a092.pdf. Online only.

Barakat, S., 2003. *Housing Reconstruction after Conflict and Disaster*. Humanitarian Practice Network, Overseas Development Institute, London.

Barakat, S. and Ellis, S., 1996. Researching under fire: issues for consideration when collecting data and information in war circumstances, with specific reference to relief and reconstruction projects. *Disasters*, 20(2). 149–156.

Corsellis, T. and Vitale, A., 2005. *Transitional Settlement – Displaced Populations*. Oxfam Publishing, Oxford.

Cuny, F., 1983. *Disasters and Development*. Oxford University Press, Oxford.

Davidson, C. H., Johnson, C., Lizarralde, G., Dikmen, N. and Sliwinski, A., 2007. Truths and myths about community participation in post-disaster housing projects. *Habitat International*, 31(1). 100–115.

Davis, I., 1977. Emergency shelter. *Disasters*, 1(1). 23–40.

Davis, I., 1978. *Shelter after Disaster*. Oxford Polytechnic Press, Oxford.

Davis, J. and Lambert, R., 2002. *Engineering in Emergencies: A Practical Guide for Relief Workers* (2nd edn). ITDG Publishing, London.

Flyvbjerg, B., 2006. Five misunderstandings about case-study research. *Qualitative Inquiry*, 12(2). 219–245.

Gall, M., 2004. Where to go? Strategic modelling of access to emergency shelters in mozambique. *Disasters*, 28(1). 82–97.

Greenomics Indonesia and WWF Indonesia, 2005. *The Implementation Design: Timber for Aceh*. Greenomics Indonesia and WWF (World Wildlife Fund), Jakarta.

Hayles, C. S., 2010. An examination of decision making in post disaster housing reconstruction. *International Journal of Disaster Resilience in the Built Environment*, 1(1). 103–122.

Johnson, C., 2007. Impacts of prefabricated temporary housing after disasters: 1999 earthquakes in Turkey. *Habitat International*, 31(1). 36–52.

Johnson, C., Lizzaralde, G. and Davidson, C. H., 2006. A systems view of temporary housing projects in post-disaster reconstruction. *Construction Management and Economics*, 24(4). 367–378.

Karunasena, G. and Rameezdeen, R., 2010. Post-disaster housing reconstruction comparative study of donor vs owner-driven approaches. *International Journal of Disaster Resilience in the Built Environment*, 1(2). 173–191.

Kelman, I., 2005. Operational ethics for disaster research. *International Journal of Mass Emergencies and Disasters*, 23(3). 141–158.

Kennedy, J., Ashmore, J., Babister, E. and Kelman, I., 2008. The meaning of 'build back better': evidence from post-tsunami aceh and Sri Lanka. *Journal of Contingencies and Crisis Management*, 16(1). 24–36.

Leon, E., Kelman, I., Kennedy, J. and Ashmore, J., 2009. Capacity building lessons from a decade of transitional settlement and shelter. *International Journal of Strategic Property Management*, 13(3). 247–265.

Lewis, J. and Mokrane, F. D., 1981. *Assistance Préparatoire pour la Reconstruction des Etablissements Humains dans la Région d'EI Asnam, Algérie*. République Algérienne Démocratique et Populaire: Commission Interministérielle pour l'Étude de la Reconstruction d'EI Asnam/Centre des Nations Unies pour les Établissements Humains (CNUH Habitat) Rapport de Mission, Algiers.

Manfield, P., Ashmore, J. and Corsellis, T., 2004. Design of humanitarian tents for use in cold climates. *Building Research and Information*, 32(5). 368–378.

Merriam, S. B., 1998. *Qualitative Research and Case Study Applications in Education*. Jossey-Bass, San Francisco.

Mwiandi, S., 2008. *Moving Beyond Relief: The Challenges of Settling Kenya's Internally Displaced*. USIPeace Briefing August 2008. United States Institute of Peace, Washington, DC.

Nigg, J. M., Barnshaw, J. and Torres, M. R., 2006. Hurricane Katrina and the flooding of New Orleans: emergent issues in sheltering and temporary housing. *The Annals of the American Academy of Political and Social Science*, 604(1). 113–128.

Oliver-Smith, A., 1986. *The Martyred City*. University of New Mexico Press, Albuquerque.

Oliver-Smith, A., 1990. Post-disaster housing reconstruction and social inequality: a challenge to policy and practice. *Disasters*, 14(1). 7–19.

Polman, L., 2010. *The Crisis Caravan, What's Wrong with Humanitarian Aid?* Henry Holt and Company, New York.

Quarantelli, E. L., 1995. Patterns of shelter and housing in US disasters. *Disaster Prevention and Management*, 4(3). 43–53.

Rand, E. C., Hirano, S. and Kelman, I., 2011. Post-tsunami housing resident satisfaction in Aceh. *International Development Planning Review*, 33(2). 187–211.

Resster, E. M., 1978. Accountability as a programme philosophy. *Disasters*, 2(2–3). 129–133.

Rosenstein, D. L., 2004. Decision-making capacity and disaster research. *Journal of Traumatic Stress*, 17(5). 373–381.

Saunders, G., 2004. Dilemmas and challenges for the shelter sector: lessons learned from the sphere revision process. *Disasters*, 28(2). 160–175.

Setchell, C., 2006. Post-crisis, long-term shelter response is vital. *Habitat Debate*, 12(4). 9.

Shelter Centre, 2010. *Shelter After Disaster*. Shelter Centre, Geneva.

Shelter Projects, 2008. *Shelter Projects 2008*. IFRC, UN-Habitat, and UNHCR, Geneva.

Shelter Projects, 2009. *Shelter Projects 2009*. IFRC and UN-Habitat, Geneva.

Shelterproject, 2003. *Report on the Transitional Settlement Sector*. Shelterproject, Cambridge.

Sheppard, S. and Hill, R., 2005. *The Economic Impact of Shelter Assistance in Post-Disaster Settings*. CHF International, Silver Spring, MD.

Silkin, T. and Hendrie, B., 1997. Research in the war zones of Eritrea and Northern Ethiopia. *Disasters*, 21(2). 166–176.

Snarr, D. N. and Brown, E. L., 1980. User satisfaction with permanent post-disaster housing: two years after hurricane Fifi in Honduras. *Disasters*, 4(1). 83–91.

Snarr, D. N. and Brown, E. L., 1982. Attrition and housing improvements: a study of post-disaster. *Disasters*, 6(2). 125–131.

Snarr, D. N. and Brown, E. L., 1994. Post-disaster housing reconstruction: a longitudinal study of resident satisfaction. *Disasters*, 18(1). 76–80.

SPHERE Project, 2004. *Humanitarian Charter and Minimum Standards in Disaster Response*. Sphere Project, Geneva.

Turner, J. F. C., 1972. Housing as a verb. *Freedom to Build: Dweller Control of the Housing Process*, J. F. C. Turner and R. Fichter (eds). Macmillan Company, New York. 148–175.

UNDRO, 1982. *Shelter after Disaster: Guidelines for Assistance*. UNDRO (United Nations Disaster Relief Organisation), New York.

UNFPA, 2007. *State of World Population 2007: Unleashing the Potential of Urban Growth*. UNFPA (United Nations Population Fund), New York.

URD, 1999. *Capitaliser sur l'evaluation des interventions liées à l'ouragan Mitch: Un outil pour la préparation à la gestion de catastrophes naturelles de grande ampleur*. URD (urgence rehabilitation développement), Paris.

URD, 2005. *12 leçons tirées de Mitch et des tremblements de terre au Salvador qui peuvent être utiles pour la réponse au tsunami*. URD (urgence rehabilitation développement), Paris.

Yin, R. K., 2003. *Case Study Research: Design and Methods*. Sage, London.

Disability and public shelter in emergencies

JOHN TWIGG[1], MARIA KETT[2], HELEN BOTTOMLEY[2], LIN TZE TAN[1] AND HUSSAM NASREDDIN[1]

[1]Department of Civil, Environmental and Geomatic Engineering, University College London, London WC1E 6BT, UK
[2]Leonard Cheshire Centre for Disability and Inclusive Development, University College London, London, UK

This paper reviews current practice regarding people with disabilities in public (or communal) emergency shelter management. It shows that provision for disabled people generally fails to meet their needs and describes the main problem areas. These problems are set in the context of management and staff failings as well as underlying weaknesses in disaster management structures as a whole. The latter include outdated attitudes towards disability, the invisibility of people with disabilities to emergency officials and relief workers and misguided assumptions about the capacities of disabled people and their organizations to manage and respond to crises. The paper concludes with suggestions about how these challenges might be overcome.

1. Introduction

There are estimated to be over 600 million people with disabilities[1] worldwide: 80 per cent of them live in developing countries, many of which are disaster-prone. Disabled people are highly vulnerable to disasters, more on account of social marginalization than impairments (Kett and Twigg, 2007). This paper presents findings of a review of current practice regarding people with disabilities in public (communal) emergency shelter management: by public emergency shelter we mean any building or facility used for the purpose of sheltering large numbers of people during and after emergencies, be it purpose-built or taken over at times of crisis. We reviewed documents on disability in emergency preparedness and response to natural hazard events; much of this was 'grey' literature from operational agencies. Only recently has a body of literature on disabled people's vulnerability to disaster, the social impact of disasters on disabled people and appropriate models of disaster management begun to emerge (Hemingway and Priestley, 2006), much of it being from the USA. Nevertheless, there is a growing and consistent body of evidence from different parts of the world showing that: people with disabilities are disproportionately affected by disasters; emergency preparedness, response and recovery operations are planned and implemented with little regard for their needs and capacities; lesson learning is weak even where problems are identified; and progress towards inclusion is very slow (HarrisInteractive/National Organization on Disability, 2004; Cordeiro and Deshpande, 2005a,b; IDRM, 2005; National Council on Disability, 2005; Kett and Twigg, 2007; Clive et al., 2010; Wisner, in press).

The subject of public emergency shelters is neglected even within the literature on disability and disasters (National Council on Disability, 2009, p. 125). Attention is given to other issues, such as warning delivery and response, evacuation from homes and public buildings, aid distribution and recovery assistance. Moreover, the discussion of emergency communal shelters and disability proceeds largely from the standpoint of medical and charity models of disability. The medical model defines disabled people individually by their impairments and offers medical or technical solutions. The charity perspective is that disabled people are merely to be pitied and helped. Disability rights' advocates press for adoption of a social model, based on human rights, which sees disability as the social consequence of having an impairment: therefore, the inequalities faced by disabled people can only be overcome if society becomes less discriminatory and more inclusive (Kett et al., 2005; Clive et al., 2010).

2. Good practice (the normative model)

The main elements of good disability practice in communal shelter management can be summed up as follows (National Council on Disability, 2005; National Organization on Disability, 2005b; Oosters, 2005; Gibson and Hayunga, 2006; US Department of Justice, 2007a,b; Federal Emergency Management Agency, 2008; Kailes, 2008; Curtin, 2009; Clive et al., 2010):

- ensuring equal access (this includes accessible parking, exterior routes, entrances, interior routes to the shelter area and toilets serving the shelter area);
- shelter staff (professional and volunteer) are trained in meeting disabled people's needs and plan for these in advance;
- availability of food, including special diets;
- provision of technical support (e.g. access to electricity for medical and mobility devices and refrigeration for medication);
- use of appropriate methods of communication (visual, audio and interpreters);
- appropriate and sufficient medical and volunteer assistance is available (including families, personal support networks and, where appropriate, care animals);
- promoting and sustaining disabled people's independence and safety when sheltering;
- involvement of people with disabilities in shelter planning;
- monitoring and evaluating shelter activity and practice;
- assistance in returning home, or provision of/ assistance in finding temporary accommodation for those unable to return to their homes immediately after an event.

3. Experiences from disasters

The literature indicates strongly that public shelter provision for disabled people in emergencies generally fails to meet their needs (Blanck, 1995; Anam et al., 2002; National Council on Disability, 2005; 2006a,b; 2009; Oosters, 2005; Freking, 2006; Gibson and Hayunga, 2006; Hemingway and Priestley, 2006; National Organization on Disability, 2006; White et al., 2007; Kailes, 2008; Clive et al., 2010; Noman Khan and Bari, undated). The following problems are widespread, throughout the world.

Access (physical):

- access routes that disabled people cannot use;
- shelters that are physically inaccessible (location or entrance facilities).

Admission:

- people with disabilities refused admission to shelters on the grounds that the shelters cannot manage their disability and they should seek specialist facilities (particularly those with mental health issues).

Needs assessment:

- registration and assessment procedures that fail to pick up special needs, particularly functional needs (in addition to the lack of preexisting special needs registries).

Facilities:

- lack of accessible restrooms, cots, toilets, bathrooms, showers.

Food:

- difficulty in getting access to meals that are provided;
- lack of options for specific dietary needs (e.g. people unable to chew, diabetics).

Medication:

- problems accessing medication, medical equipment and medical supplies;
- problems with assistance in refrigerating medication.

Communications:

- telephone and other communications equipment inaccessible;
- lack of or inadequate signage, captioning and translation in communicating messages.

Personal support:

- family members, carers and members of personal support networks with skills in supporting people with functional needs discouraged from accompanying them into shelters or denied access;
- no provision made for service animals, leading to their exclusion from shelters;
- marginalization and discrimination by other shelter users, particularly in crowded shelters.

The harmful effects of such practices were frequently exemplified during Hurricane Katrina in the USA in 2005. Regarding access to medical facilities, for instance, a volunteer in one shelter recorded that:

> I have seen many frail people struggle to climb or descend the stairs in order to get medical attention, and I have personally seen two very exhausted men in wheelchairs almost decide to forego triage or other medical attention because of the difficulty of accessing this unit (National Council on Disability, 2006a).

Elsewhere, inadequate communication marginalized deaf evacuees in accessing support services:

> Because the majority of the deaf evacuees used American Sign Language, few of the shelter volunteers could communicate with them . . . The deaf evacuees missed out on important announcements that streamed through the public address systems of the large shelters. The announcements instructed evacuees which line to stand in to get federal financial assistance, which line to go to for medical appointments, and how to register to locate missing relatives. Deaf evacuees never got this information and thus were passed over for services the first week at the shelter (White, 2006, pp. 50–51).

Disabled people's separation from their families and support networks resulted in:

> some people being . . . transferred unnecessarily to medical shelters or nursing homes. Others were not identified because of the lack of trained eyes as well as the lack of or inadequate screening questions. This caused some individuals' conditions to deteriorate to the point that they did require transfer to a hospital, nursing home, or medical shelter (National Organization on Disability, 2006, p. 9).

US research also suggests that lack of transportation, failure of a 'buddy' (personal support) system and the belief that a shelter will be unable to accommodate their specific needs make people with disabilities reluctant to evacuate in response to warnings (National Council on Disability, 2009, p. 129). Where evacuation is more frequent, in response to seasonal hazards, previous unpleasant experience of evacuation shelters may make disabled people reluctant to venture into them again. In Bangladesh, inaccessibility, lack of safety and poor sanitation facilities in cyclone shelters, combined with cultural constraints, deter disabled people, especially women, from taking refuge (Anam et al., 2002, pp. 29–30).

4. Contextualizing shelter management and its failings

Most of the problems listed above are directly attributable to limitations in staffing and management. The literature already cited also demonstrates:

- an acute shortage of shelter personnel trained in planning for people with disabilities, working with them and dealing with their needs;
- lack of co-ordination and communication between emergency managers and between shelters regarding special needs, referrals and acquisition of additional specialist human and material resources;
- failure to survey shelter sites with accessible features in mind; staff not trained to assess shelters for accessibility;
- lack of engagement with disabled people's organizations (DPOs) and other specialist disability organizations, both in disaster planning and during events, to understand disabled people's needs and provide support.

These management and staffing failings reflect underlying weaknesses in disaster management structures as a whole. Many studies identify poor planning for people with disabilities in disasters, confusion about the roles and responsibilities of care giving and emergency management staff and misplaced assumptions among emergency managers that disability service providers can meet all disabled people's needs in a crisis (Clive et al., 2010). In its review of performance in Hurricane Katrina, the US National Organization on Disability concluded that 'All levels of government experienced systemic failures in their efforts to respond to the needs of the disability and aging populations.' (National Organization on Disability, 2006, p. 16), while the National Council on Disability noted that 'many of the problems ... are systemic in nature and were not caused solely by the hurricanes. The challenges faced by people with disabilities during and after the hurricanes, while unique in scope and proportion, were similar to the challenges people with disabilities face on a day-to-day basis' (National Council on Disability, 2006a).

The scale of this institutional challenge is indicated by the repeated failure to learn lessons from past experience. According to the US National Council on Disability:

> Many of these barriers are not new. Information and lessons learned are not shared across agency lines, and thus experience does not enlighten the development of new practices. Many accessibility lessons learned during previous disasters are not incorporated in subsequent planning, preparedness, response, and recovery activities ... Lessons learned after each disaster about physical, communication, and program access for recovery centers, and other structures and buildings used in connection with disaster operations (e.g. first aid stations, mass feeding areas, portable payphone stations, portable toilets, and temporary housing, as well as shelter identification, access, management, training, and services) do not appear to get integrated into subsequent practice (National Council on Disability, 2005, pp. 11, 42).

Deeper attitudinal problems underpin this reluctance to change. One is the persistence of the medical model of disability in emergency management, whereby people with disabilities are seen purely in terms of their specific disabilities. This causes them to be barred from many public shelters on the misguided assumption that they need specialist support and facilities that are not available in the shelters (e.g. National Council on Disability, 2005, p. 19; National Organization on Disability, 2006, p. 13). It also leads to them being separated from their support networks and communities. A more effective approach, now being advocated, is to use a flexible framework addressing a broad set of function-based needs (e.g. communication, medical, maintaining functional independence, supervision and transportation) and reflecting the capabilities of the individual, irrespective of any specific diagnosis or status (Kailes and Enders, 2007; Federal Emergency Management Agency, 2008; Martin,

2009; National Council on Disability, 2009, pp. 316–317). Adoption of this approach could lead to 'a paradigm shift in emergency planning in that it fosters the development of an operational set of predictable supports' (Parsons and Fulma, 2007).

Disabled people remain largely invisible to emergency officials and relief workers. This is revealed, for example, by the view commonly held by emergency staff that there are very few people with disabilities in a disaster-affected area even when it should be obvious that there are many more (Kett et al., 2005, pp. 13, 17). As US disability and disaster specialist June Kailes points out, 'Most disaster response systems are designed for people who can walk, run, see, drive, read, hear, speak and quickly understand and respond to instructions and alerts' (Kailes, 2008, p. 10). Even basic information on numbers and location of disabled people is likely to be missing or inadequate (Kett and Twigg, 2007). A study of community organizations and public safety agencies in California, in 1984, revealed that 'relatively little activity had taken place' in emergency planning for disabled people (Tierney et al., 1988, pp. 84–88). Interviews with local emergency managers in Manila, Mumbai and Mexico City, carried out for a United Nations University study some years ago, indicated that disabled people were not even identified specifically as a vulnerable group (Wisner, in press). After the 9/11 attacks in New York, the Center for Independence of the Disabled reported that 'people at the Red Cross were polite and interested, but everything had to be brought to their attention.' (National Council on Disability, 2005, p. 40). Research after the Indian Ocean Tsunami of 2004 concluded that relief agencies' 'awareness [of disabled people and their needs] was low, with disability not being a topic of discussion at the majority of organizing and co-ordinating meetings' (IDRM, 2005, p. 6); and few relief agencies included disability in their evaluation methodologies (Hemingway and Priestley, 2006, p. 61). In Bangladesh, the variance between official assertions and community views of the level of assistance given to people with disabilities in preparing for disasters and accessing shelters (Anam et al., 2002, p. 32) may be due in part to officials' inability to perceive and understand disability in the community.

If recognized at all, people with disabilities are likely to be put with others into the catch-all category of vulnerable groups, which fails to identify their specific needs. It is not that emergency managers overlook disability alone: there is plenty of evidence to suggest that in many areas of crisis management and disaster response, including sheltering, there are widespread failures to recognize the specific needs of other marginalized and vulnerable groups (e.g. Phillips et al. (eds), 2010). Even when special needs are identified, there may be inconsistency between different emergency organizations in defining such groups and failure to appreciate that disability takes different forms and hence leads to different needs. Oversights and omissions appear almost inevitable in such circumstances (White et al., 2007, pp. 28–29).

With such attitudinal barriers, it is not surprising that disabled people are usually seen as mere passive recipients of aid. DPOs and other organizations representing or supporting them are excluded from planning and implementing emergency response. Researchers in India after the 2004 tsunami found that 'inclusion of disabled people seems to be limited to surveys, receiving relief and aids and equipment, and does not involve inclusion in planning, decision-making or management' (Kett et al., 2005, p. 6). In a US survey after Hurricane Katrina, five out of six emergency managers interviewed reported having a plan in place that included people with disabilities, but only two stated that disability organizations had been involved in the preparation of the plan (White et al., 2007, p. 15). During Katrina, workers in Centers for Independent Living complained that they were denied access to shelters, that emergency staff were uninterested in or unaware of the skills they had to offer or, in some cases, that they were simply 'dumped on' by emergency managers, who referred all disabled people and disability problems to them (White et al., 2007, pp. 22–23). Even on disability questions, emergency managers are more likely to communicate with other government agencies than with disability organizations, who are kept out of the loop (e.g. White et al., 2007, pp. 18, 22). Much of the communication received by the New Orleans Center for Independent Living in the Katrina disaster was from other disability organizations out-of-state; this resulted in tangible support in the form of money, volunteers, durable medical equipment and clothing. 'The overwhelming response to the question "What type of resources would be most helpful to better serve people they serve in an emergency or disaster?" was the need for better communication and networking at all levels' (White et al., 2007, p. 21).

Agencies that support inclusion, in principle, may not understand its implications in practice (Kett et al., 2005, pp. 26–32). This is not just an accidental feature of disaster response but also a fundamental institutional failing. After the experiences of the 2004 Indian Ocean tsunami, it was claimed that 'Many disabled people have been and are being structurally excluded from relief programmes. Their basic needs are simply not adequately met and their human rights are at best ignored, at worst abused' (Oosters, 2005, p. 8).

5. Ways forward

A wide range of complementary efforts will be required to overcome these problems. These are outlined in the following paragraphs. Here we are principally concerned with solutions to the underlying institutional and other causes of poor shelter facilities and management, rather than with specific technical matters.

There is no shortage of technical guidance on public emergency shelter design although its quality and consistency vary greatly. In Australia and the USA, for example, it is thorough, incorporating a range of disability-related provisions, many of them straightforward (Rossiter and Robinson, 2005; Mullins Consulting, 2006; US Department of Justice, 2007a,b; Nasreddin, 2010). Improved guidance is also becoming available in developing countries (e.g. Handicap International Bangladesh, 2005; Hans et al., 2005; Curtin, 2009). But even recent guidance can be far from sufficient. Official guidelines for the design and construction of cyclone and tsunami shelters published by the Government of India and United Nations Development Programme in 2006, when considering access arrangements and staircases, suggest merely that: 'a ramp with a slope of 1:8 to 1:10 may be considered up to the first floor for carrying physically disabled and elderly people' (Arya and Agarwal, 2006, p. 10), making no mention of other forms of provision for people with disabilities within the shelters.

Examples of successful sheltering demonstrate that good organization and planning are key elements in shelter access and provision for disabled people, at strategic and operational levels. This was certainly the case in evacuations of residents of nursing homes in Florida to avoid Hurricane Charley in 2004 (Kuba et al., 2004), although helped by the fact that these were people already living in institutional environments rather than independently. Duval County, Florida, has implemented an 'Adopt-A-Shelter' programme in partnership with local hospitals and medical supply companies to ensure that special needs shelters are fully stocked prior to disasters, to develop an inventory for these shelters and to provide resources in the event of an emergency (National Council on Disability, 2006a). The Senior Safe Center programme, in place in the states of Alabama and Florida since 2008, involves building and retrofitting dual-purpose centres and shelters for senior citizens (Clive et al., 2010, p. 202). There have been many calls for better registries of disabled people's locations and needs and of the material, financial and human resources that can be called upon to support them in a disaster (e.g. White et al., 2007, pp. 35–36). In the USA new online networking tools are being developed to support this, such as the Special Population Information Registry designed by the Center for Disaster Risk

Policy at Florida State University (www.spinreg.org). There have also been suggestions that geographical information systems technology can be applied to this task (Enders and Brandt, 2007). In Kobe, Japan, a detailed survey and mapping project has been undertaken to identify disabled people living in hazard-prone locations and to raise awareness of their situation and needs in an emergency with disaster planners and among their communities (Tatsuki and Comafay, 2010).

However, addressing underlying institutional weaknesses is essential to achieve widespread and lasting changes in operational practice. The most commonly advocated approach is to sensitize emergency management organizations and their staff through training, guidelines and other technical resources. This is how the Inter-Agency Standing Committee, the main mechanism for co-ordinating international humanitarian assistance, recommends improvement of services to older people, many of whom may have some disability (Inter-Agency Standing Committee, 2008, p. 6). In the USA it has been suggested that such training should be made a pre-requisite for career progression in emergency management (Fox et al., 2007, p. 203).

Such standard prescriptions are necessary but insufficient because they do not address the fundamental principle of inclusion, a point made firmly by the US National Council on Disability after Hurricane Katrina: 'It is important to recognize that some of the challenges faced by people with disabilities who sought assistance in shelters are inherent in any disaster response – the initial general confusion, an inadequate number of trained personnel, etc. However many of the most significant problems could have been avoided with more inclusive emergency planning' (Nation Council on Disability, 2006a). Similarly, the US National Organization on Disability identified several exemplary general population shelters that were accessible to people with disabilities during Katrina. One city mayor designated a convention centre as a general shelter and ensured that it included interpreters for the deaf, accessible shuttle services, internet access, employment opportunities and information on how to find accessible housing. The shelter operators were not experienced emergency managers; rather, the key to their success was their attitude of inclusiveness (National Organization on Disability, 2006).

Legislation regarding equality or requiring providers of public goods and services to ensure that these are accessible to people with disabilities may be an important driver of change in some countries. This is notably the case in the USA where the Americans with Disabilities Act (1990) has stimulated several Federal Government agencies to produce guidance relating to disability and emergency management (e.g. US Department of Justice, undated; US Department of Transportation, 2003). The Act also requires shelters to conduct operations in ways that offer people with disabilities the same benefits (such as safety, comfort, food, medical care, and the support of family and friends) that are provided to people without disabilities (US Department of Justice, 2007a,b; Federal Emergency Management Agency, 2008, p. 25; Kailes, 2008, p. 69). Further legislative and administrative improvements followed hurricanes Katrina and Rita in 2005 (Parsons and Fulma, 2007; Clive et al., 2010, p. 195). County emergency managers surveyed in the USA identified federal government mandates as the single most important reason for modifying county disaster plans, with prompting by people with disabilities being a very minor factor in driving change (Fox et al., 2007, p. 200).

However, it cannot be assumed that legislative change will translate easily and quickly into change on the ground. It was some years before the Americans with Disabilities Act began to make its mark on formal guidelines, while the University of Kansas 'Nobody Left Behind' study (2003–5) found that only 27 per cent of the county emergency managers surveyed had completed the Federal Emergency Management Agency's Emergency Planning and Special Populations course and only 20 per cent had specific guidelines in place to help people with disabilities in disasters (Fox et al., 2007, pp. 201–202). In India the Persons with Disabilities Act 1996 was said after the 2004 tsunami to have 'failed to produce a noticeable change in society', with 'slow implementation of its provisions'. Planning and designs for post-tsunami housing and reconstruction programmes there seem largely to have ignored access and other disability issues (IDRM, 2005, pp. 17, 23–25, 32–35). A similar situation was encountered in Indonesia, where the provisions of the 1997 Act Concerning People with Disabilities were not echoed in pre-tsunami disaster plans and procedures and little attention was paid to disabled people in post-tsunami disaster planning (IDRM, 2005, pp. 42–54).

More specific legal provisions or standards may be required to bring about measurable changes on the ground in areas such as access arrangements and provision of accessible toilets and bathrooms (e.g. Kailes, 2008, pp. 70–71). The issue of appropriate disability standards for emergency preparedness is currently attracting attention and discussion in the USA, for instance, although the emphasis is on emergency communication and evacuation issues (American National Standards Institute, 2009). Standard setting is potentially valuable in improving shelter management, but there have been warnings about the needs for disabled people's participation in these processes; for clear expectations about what standards can deliver; and about ensuring that they are seen as issues for the whole community, not merely the disabled, to address (American National Standards Institute, 2009). Internationally, the latest revision of the Sphere Project's Humanitarian Charter and Minimum Standards in Disaster Response (Sphere Project, 2010), due to be published in 2011, is taking disability more seriously as a cross-cutting issue although the standards do not contain any specific standards for people with disabilities in communal emergency shelters (Kett, 2010).

Universal standard setting is contentious, since resources and capacity are so unevenly distributed. A complaint made in a US National Organization on Disability report regarding communications with deaf or hard-of-hearing evacuees from Katrina – that less than 30 per cent of shelters had access to American Sign Language interpreters, 80 per cent did not have text telephones, and 60 per cent did not have televisions with open caption capability (National Organization on Disability, 2006, pp. 8–9) – implies that provision of such services should be easily attainable, a notion that would be considered idealistic in many countries. Even the seemingly more down-to-earth approach of posting written announcements for people who are deaf or hard of hearing has limited value where literacy levels are low.

Moreover, the issue of disability standards in emergencies ought not to be separated from the application of disability standards in society at large. Since many public emergency shelters are not purpose-built but other public or private buildings (e.g. schools, community centres, sports halls) taken over in a crisis, their design and daily use should already be governed by appropriate regulations on such matters as inclusive access and signage, for which the legal responsibility rests with their regular owners and operators. Where this is lacking, advocacy for improved standards or compliance cannot focus on emergencies alone but has to target this wider context.

Adoption of rights-based approaches to disability in disasters has been advocated as a key to overcoming many of the problems faced by people with disabilities, in the longer term. It has been argued that 'the human rights of persons with disabilities should receive priority and that each individual has a right to a basic level of public safety. The acceptance of the existence of such a right carries with it the moral duty to ensure that this right is manifested and protected in disaster mitigation policy' (Parr, 1997, p. 2). Furthermore, 'Addressing fundamental issues of rights, inclusion and equality are effective ways of tackling vulnerability in itself. Disabled people may require many of the same support structures, aid and interventions as may other sections of the population... For example, the basic needs in immediate post-emergency phases are the same for all – water, shelter, food, sanitation, etc. It is how they are provided that makes the difference' (Kett, 2007, p. 160).

Such advocates have taken heart from the UN Convention on the Rights of Persons with Disabilities (2008), which is based on the social model of disability. Its aim is 'to promote, protect and ensure the full and equal enjoyment of all human rights and fundamental freedoms by all persons with disabilities, and to promote respect for their inherent dignity' and it calls on signatories to 'adopt all appropriate legislative, administrative and other measures' for the implementation of these rights. Its wide-ranging provisions include (Article 11) 'all necessary measures to ensure the protection and safety of persons with disabilities in situations of risk, including situations of armed conflict, humanitarian emergencies and the occurrence of natural disasters' (United Nations General Assembly, 2006, pp. 7–8, 12). The Convention has sparked off some serious discussions about inclusion and mainstreaming of disability into sustainable development (e.g. Yousif and Shaw (eds), 2008), but there is little indication yet of serious penetration into humanitarian assistance thinking and planning. While it will support the promotion of the protection and full and equal enjoyment of all human rights and fundamental freedoms by all persons with disabilities, it should not be seen as a panacea for the eradication of discrimination and inequality. Much relies on how the Convention will be implemented, monitored and evaluated at the national level (Kett et al., 2009).

Disabled people and their organizations have insisted, repeatedly, on their right to be active stakeholders in emergency planning and action, a point made emphatically in the 2007 'Verona Charter':

> Persons with disabilities and their organisations need to be actively involved in decision-making processes concerning situations of humanitarian emergencies and the occurrence of natural and man made disasters and in all the related emergency management activities. This involvement should be fostered by the development of inclusive policies at all levels starting from organisations of persons with disabilities and families, communities up to national and international organisations/bodies (Verona Charter, 2007).

In the USA guidance has been available to disabled people for some time on how to prepare for emergencies, including creating personal support networks, communicating and liaising with disaster management staff and becoming involved in emergency planning processes (American Red Cross, 1997; Kailes, 2002). Such guidance is now also recommending that people with disabilities consider how their needs might be met in public shelters (National Organization on Disability, 2005a). Nevertheless, there is evidence that disabled people and disability organizations or support groups are often not sufficiently prepared for emergencies (White et al., 2007, pp. 14–15, 29–31). There needs to be more investigation and testing of appropriate processes for liaison and negotiation between people with disabilities and shelter staff.

The need for disabled people and their organizations to be more involved in emergency planning as a whole, which is a key component of disability organizations' advocacy on disasters (Kett et al., 2005; National Organization on Disability, 2005b, 2006; Parsons and Fulma, 2007; Kailes, 2008; Wisner, in press), is also beginning to be acknowledged by policy makers (National Council on Disability, 2005; US Department of Homeland Security, 2005, p. 38). This requires the creation of networks and partnerships at all levels well in advance of emergency events, as recognized in recent guidance by the Federal Emergency Management Agency in the USA: 'emergency managers should draw from the skills and resources within special needs planning networks' (Federal Emergency Management Agency, 2008, p. 25); and it should lead to the deployment of community and other organizations in support of official agencies during events.

Many DPOs and disability organizations are involved in emergency response and relief provision to meet needs and fill gaps, but not necessarily in a planned and co-ordinated way or within the mainstream disaster assistance system. Global disability networks were quick to react with funding and material support after the 2004 tsunami and there was a similar prompt and well-networked reaction by US disability organizations after Katrina (Hemingway and Priestley, 2006, p. 63). In Katrina, US non-profit disability organizations stepped in to provide disabled people in shelters with support resources that the shelters lacked (e.g. teletypewriters, wheelchairs, walkers, oxygen) (National Council on Disability, 2006a), while a deaf people's church which provided food, clothing, children's toys, household items, hearing aids and other equipment to deaf evacuees in the Houston Astrodome 'essentially became a social agency and a center for social support for the evacuees' (White, 2006, p. 51). A decade earlier, a group in Los Angeles called Disabled People and Disaster Planning (DP2) had produced guidance arising from the experiences of the 1994 Northridge earthquake; this included advice to shelter managers on working with people with disabilities and making shelters more accessible (Wisner, in press).

Nevertheless, there is still little linking between civil society organizations representing or working for disabled people and emergency management systems. This has led to the call for efforts 'to raise the awareness of persons with disabilities to the hazards of not being proactively involved in disaster planning at both the personal and community levels. Emergency managers must be willing to reach out to the disability community more aggressively, but it is incumbent on members of the disability community to also reach out to emergency managers if they ignore this opportunity' (Fox et al., 2007, p. 203). In a US study after Hurricane Katrina, 85.7 per cent of 'disability and aging specific organizations... answered that they did not know how to link with the emergency management system' (National Organization on Disability, 2006, p. 7), echoing the findings of an earlier study which found that 53 per cent of people with disabilities surveyed did not know whom to contact regarding emergency plans for their community (National Organization on Disability, 2005b, p. 7). General guidance on building such links is starting to become available (e.g. National Organization on Disability, 2005b) and in the USA a few instances of such partnerships have been identified (e.g. National Council on Disability, 2005, pp. 46–54; 2009, pp. 299–300). Since Katrina, special needs planning groups (comprising disability and ageing organizations, other voluntary organizations working in disasters, health care providers, emergency management and first-response agencies and other institutions such as housing agencies) have begun to form at federal, state and local levels (Clive et al., 2010, pp. 206–207). In displaced persons' camps in Sudan, HelpAge International has set up older people's committees to carry out needs assessments, liaise with other agencies, share information and monitor and assist in distribution of food and other essentials (Wells, 2005, p. 10). An initiative being tested in California is the formation of Functional Assessment and Service Teams, comprising government and NGO volunteers with knowledge and understanding of disability communities, often themselves people with disabilities, who can go into shelters to assess unidentified or unmet needs, solve problems rapidly through provision of materials and services, negotiate with emergency staff and assist the transition back into communal living (Kailes, 2008, pp. 77–84). In India, establishment of targeted disaster advocacy, outreach and information services has been recommended as a key element of disaster response strategy, to distribute information to vulnerable people, identify the needs of the older and disabled population, collate evidence and assist disabled people to obtain access to relief and recovery services (SEEDS, 2001, pp. 50–53). Further research into the effectiveness of these partnerships, the processes involved in creating and maintaining them and the factors that stimulate or inhibit partnerships, in different contexts, would improve our understanding of how to make progress. One factor that is likely to be significant was identified in a US study, which, in recommending measures to improve services and facilities had to acknowledge that 'many of them will require the investment of the scarcest resource available: time on the part of already over-burdened staff of CILs [Centers for Independent Living] and local emergency management agencies' (White et al., 2007, p. 27).

Representation and leadership are critical issues. One of the messages from a recent workshop on emergency planning for disabled people organized by the American National Standards Institute was the need for 'ensuring adequate representation in leadership roles' (American National Standards Institute, 2009, pp. 2–3). A report for the National Institute on Disability and Rehabilitation Research after Hurricane Katrina went further, arguing that the existing network of Centers for Independent Living, should be given a 'leadership role in a process of bringing together disability organizations including CILs, as well as state and local emergency planners, to develop mechanisms to increase information sharing, coordination, and the development of disaster preparation and emergency response plans that incorporate people with disabilities', by such means as forming planning committees, working with emergency agencies to improve their plans, providing staff to act as contact points during emergencies and stimulating local disability organizations to engage (White et al., 2007, p. 28). One recommendation after Katrina was that parallel teams specializing in issues of disability and ageing should be put in place to work alongside the decision-making structure set out in the US National Response Plan (National Organization on Disability, 2006, p. 10).

This implies action by all relevant stakeholders, or in the words of the voluntary Verona Charter on the rescue of persons with disabilities: 'The safety of persons with disabilities in situations of humanitarian emergencies and the occurrence of natural and man made disasters is the responsibility of organizations of persons with disabilities, single individuals, public institutions at all levels, business and civil society, social partners, non-governmental organizations, educational institutions, health authorities and civil protection organizations.' (Verona Charter, 2007). To judge from the documented experiences of people with disabilities in emergency evacuation shelters, this goal is still a long way off.

Note

1. In this paper we use the terms 'disabled people' and 'people with disabilities' interchangeably although there is some debate among the disability community about which is more appropriate; the latter term is more commonly accepted in North America.

References

American National Standards Institute, 2009. *Final Workshop Report: Emergency Planning for Persons with Disabilities and Special Needs*. American Standards Institute, Washington, DC, USA.

American Red Cross, 1997. *Disaster Preparedness for People with Disabilities*. American Red Cross, Washington, DC, USA.

Anam, N., Noman Khan, A. D. M., Bari, N. and Alam, K. J., 2002. *Unveiling Darkness: The Situation Analysis on Disaster and Disability Issues in the Coastal Belt of Bangladesh*. Centre for Services and Information on Disability, Dhaka, unpublished paper.

Arya, A. S. and Agarwal, A., 2006. *Guidelines for Design and Construction of Cyclone/Tsunami Shelter*. Ministry of Home Affairs. Government of India – UNDP Disaster Risk Management Programme, New Delhi.

Blanck, P. D., 1995. *Disaster Mitigation for Persons with Disabilities: Fostering a New Dialogue*. Washington Annenberg Program, Washington, DC, USA.

Clive, A., Davis, E. A., Hansen, R. and Mincin, J., 2010. Disability. *Social Vulnerability to Disasters*, B. D. Phillips, S. K. Thomas, A. Fothergill and I. Blinn-Pyke (eds). CRC Press, Boca Raton. 187–216

Cordeiro, V. J. and Deshpande, S. S., 2005a. *Research Report on Disability and the Tsunami in India. Part I.* International Disability and Development Consortium, Brussels.

Cordeiro, V. J. and Deshpande, S. S., 2005b. *Research Report on Disability and the Tsunami in India. Part II.* International Disability and Development Consortium, Brussels.

Curtin, M., 2009. *Guidelines for Creating Barrier-free Emergency Shelters*. Handicap International Nepal, Kathmandu.

Enders, A. and Brandt, Z., 2007. Using geographical information system technology to improve emergency management and disaster response for people with disabilities. *Journal of Disability Policy Studies*, 17. 223–229.

Federal Emergency Management Agency, 2008. *Interim Emergency Management Planning Guide for Special Needs Populations*. Federal Emergency Management Agency, Washington, DC, USA.

Fox, M. H., White, G. W., Rooney, C. and Rowland, J. L., 2007. Disaster preparedness and response for persons with mobility impairments: results from the University of Kansas nobody left behind study. *Journal of Disability Policy Studies*, 17. 196–205.

Freking, K., 2006. *Mentally Disabled Faced Bias Post-Katrina*. Associated Press, Washington, DC, USA. www.ncd.gov/newsroom/inthenews/disabled_faced_07-14-06.htm (accessed online 30 November 2006).

Gibson, M. J. and Hayunga, M., 2006. *We Can do Better: Lessons Learned for Protecting Older Persons in Disasters*. AARP Public Policy Institute, Washington, DC, USA.

Handicap International Bangladesh, 2005. *How to Include Disability Issues in Disaster Management, following Floods 2004 in Bangladesh*. Handicap International Bangladesh, Dhaka.

Hans, A., Sharma, R. K., Mohanty, R. and Mishra, P., 2005. *Training Manual for Inclusion of Disability in Disaster Response*. Shanta Memorial Rehabilitation Centre, Bhubaneswar.

HarrisInteractive/National Organization on Disability, 2004. *Emergency Preparedness Survey: Final Report*. HarrisInteractive/National Organization on Disability, Washington, DC, USA.

Hemingway, L. and Priestley, M., 2006. Natural hazards, human vulnerability and disabling societies: a disaster for disabled people?. *Review of Disability Studies*, 2(3). 57–67.

Inter-Agency Standing Committee, 2008. *Humanitarian Action and Older Persons: An Essential Brief for Humanitarian Actors*. Inter-Agency Standing Committee, Geneva.

IDRM, 2005. *Disability and Early Tsunami Relief Efforts in India, Indonesia and Thailand*. International Disability Rights Monitor/Center for International Rehabilitation, Washington, DC, USA.

Kailes, J. I., 2002. *Emergency Evacuation Preparedness: taking Responsibility for your Safety. A Guide for People with Disabilities and other Activity Limitations*. Center for Disability and the Health Professions, Western University of Health Sciences, Pomona, CA.

Kailes, J. I., 2008. *Southern California Wildfires After Action Report*. Center for Disability and the Health Professions, Western University of Health Sciences, Pomona, CA.

Kailes, J. I. and Enders, A., 2007. Moving beyond 'special needs': a function-based framework for emergency management and planning. *Journal of Disability Policy Studies*, 17. 230–237.

Kett, M., 2007. Conflict recovery. *Disability and Inclusive Development*, T. Barron and P. Amerena (eds). Leonard Cheshire International, London. 155–186.

Kett, M., 2010. Personal communication from Dr Maria Kett, Focal Point for Disability in the Sphere Project Handbook Revision 2011.

Kett, M. and Twigg, J., 2007. Disability and Disasters: Towards an Inclusive Approach. *World Disasters Report 2007: Focus on Discrimination*. International Federation of Red Cross and Red Crescent Societies, Geneva. 87–109.

Kett, M., Stubbs, S. and Yeo, R., 2005. *Disability in Conflict and Emergency Situations: Focus on Tsunami-affected Areas*. International Disability and Development Consortium, Brussels.

Kett, M., Lang, R. and Trani, J.-F., 2009. Disability, development and the dawning of a new convention: a cause for optimism?. *Journal of International Development*, 21(5). 649–661.

Kuba, M., Dorian, A., Kuljian, S. and Shoaf, K., 2004. *Elderly Populations in Disasters: Recounting Evacuation Processes from Two Skilled-Care Facilities in Central Florida, August 2004*. Natural Hazards Center, Boulder, Quick Response Report 172.

Martin, K., 2009. *A Shared Responsibility: The Need for an Inclusive Approach to Emergency Planning for People with Disabilities*. Coalition of People with Disabilities, Vancouver, BC.

Mullins Consulting, 2006. *Design Guidelines for Queensland Public Cyclone Shelters*. Queensland Government Department of Public Works, Brisbane.

Nasreddin, H., 2010. *Literature Review on Structural Design of Shelters for People with Disabilities*. University College London, Department of Civil, Environmental and Geomatic Engineering, unpublished report.

National Council on Disability, 2005. *Saving Lives: Including People with Disabilities in Emergency Planning*. National Council on Disability, Washington, DC, USA.

National Council on Disability, 2006a. *The Impact of Hurricanes Katrina and Rita on Persons with Disabilities: A Look Back and Remaining Challenges*. National Council on Disability, Washington, DC, USA. www.ncd.gov/newsroom/publications/2006/hurricanes_impact.htm (accessed 10 May 2010).

National Council on Disability, 2006b. *The Needs of People with Psychiatric Disabilities during and after Hurricanes Katrina and Rita: Position Paper and Recommendations*. National Council on Disability, Washington, DC, USA. www.ncd.gov/newsroom/publications/2006/peopleneeds.htm (accessed 28 May 2010).

National Council on Disability, 2009. *Effective Emergency Management: Making Improvements for Communities and People with Disabilities*. National Council on Disability, Washington, DC, USA.

National Organization on Disability, 2005a. *Prepare Yourself: Disaster Readiness Tips for People with Disabilities*. National Organization on Disability, Washington, DC, USA.

National Organization on Disability, 2005b. *Guide on the Special Needs of People with Disabilities for Emergency Managers, Planners and Responders*. National Organization on Disability, Washington, DC, USA.

National Organization on Disability, 2006. *Report on Special Needs Assessment for Katrina Evacuees (SNAKE) Project*. National Organization on Disability, Washington, DC, USA.

Noman Khan, A. H. M. and Bari, N., undated, *Disabled People in Flood Situation: Suggested Interventions for Post Flood Rehabilitation of Disabled People*. National Forum of Organizations Working with the Disabled, Dhaka.

Oosters, B., 2005. *Looking with a Disability Lens at the Disaster Caused by the Tsunami in South-East Asia*. CBM International, Brussels.

Parr, A., 1997. Disasters and human rights of persons with disabilities: a case for an ethical disaster mitigation policy. *Australian Journal of Emergency Management*, 12(4). 2–5.

Parsons, B. S. and Fulma, D., 2007. *The Paradigm Shift in Planning for Special-Needs Populations*. Paper to 2007 FEMA Emergency Management Higher Education Conference, Washington, DC, USA, Interagency Coordinating Council on Emergency Preparedness and Individuals with Disabilities. www.disabilitypreparedness.gov/paradigm.htm (accessed 21 May 2010).

Phillips, B. D., Thomas, S. K., Fothergill, A. and Blinn-Pike, L. (eds), 2010. *Social Vulnerability to Disasters*. CRC Press, Boca Raton.

Rossiter, J. L. and Robinson, D. M., 2005. *Design guidelines for Australian Public Cyclone Shelters*. Paper to International Symposium on Disaster Reduction on Coasts, Monash University, Melbourne, 14–16 November.

SEEDS, 2001. *High Powered Committee on Disaster Management Report on Women, Children, Aged and Challenged in Disaster*. Government of India, New Delhi.

Sphere Project, 2010. www.sphereproject.org/component/option,com_frontpage/Itemid,200/lang,english/ (accessed 30 June 2010).

Tatsuki, S. and Comafay, N., 2010. *Evacuation and sheltering assistance planning for special needs population: Kobe Disadvantaged Population Mapping Project*, Paper to International Sociological Association World Congress of Sociology, Gothenburg, Sweden, July 15, 2010.

Tierney, K. J., Petak, W. J. and Hahn, H., 1988. *Disabled Persons and Earthquake Hazards*. University of Colorado Institute of Behavioral Science, Boulder.

United Nations General Assembly, 2006. *Final Report of the Ad Hoc Committee on a Comprehensive and Integral International Convention on the Protection and Promotion of the Rights and Dignity of Persons with Disabilities: Note by the Secretary-General*. New York, United Nations General Assembly A/6/611. http://daccess-dds-ny.un.org/doc/UNDOC/LTD/N06/645/30/PDF/N0664530.pdf?OpenElement (accessed 30 June 2010).

US Department of Homeland Security, 2005. *Individuals with Disabilities in Emergency Preparedness: Executive Order 13347. Annual Report July 2005*. Department of Homeland Security, Washington, DC, USA.

US Department of Justice, undated. *An ADA Guide for Local Governments: Making Community Emergency Preparedness and Response Programs Accessible to People with Disabilities*. Department of Justice, Washington, DC, USA.

US Department of Justice, 2007a. The ADA and emergency shelters: access for all in emergencies and disasters. www.ada.gov/pcatoolkit/chap7shelterprog.pdf (accessed 1 December 2009).

US Department of Justice, 2007b. ADA checklist for emergency shelters. www.ada.gov/pcatoolkit/chap7shelterchk.htm (accessed 1 December 2009).

US Department of Transportation, 2003. *Emergency Preparedness Guidelines for People with Disabilities*. Department of Transportation, Washington, DC, USA.

Verona Charter, 2007. *"Verona Charter" on the rescue of persons with disabilities in case of disasters: declaration of the participants in the Consensus Conference held in Verona on 8–9 November 2007*. www.internazionali.ulss20.verona.it/docs/projects/rdd/veronacharter.pdf (accessed 23 March 2010).

Wells, J., 2005. *Protecting and Assisting Older People in Emergencies*. Network Paper 53, Humanitarian Practice Network, London.

White, B., 2006. Disaster relief for deaf persons: lessons from hurricanes Katrina and Rita. *Review of Disability Studies*, 2(3). 49–56.

White, G. W., Fox, M. H., Rooney, C. and Cahill, A., 2007. *Assessing the Impact of Hurricane Katrina on Persons with Disabilities*, University of Kansas, Center on Independent Living/National Institute on Disability and Rehabilitation Research.

Wisner, B. in press. Disability and disaster: victimhood and agency in earthquake risk reduction. *Earthquakes*, C. Rodrigue and E. Rovai (eds). Routledge, London. Draft at www.radixonline.org/disability.html (accessed 30 November 2009).

Yousif, H. M. and Shaw, D. (eds)., 2008. *UN Convention on the Rights of Persons with Disabilities: A Call for Action on Poverty, Discrimination and Lack of Access*. Economic Commission for Africa/Leonard Cheshire Disability, Addis Ababa/London.

Index

References in **Bold** represent illustrations.

For Product Safety Concerns and Information please contact our EU
representative GPSR@taylorandfrancis.com
Taylor & Francis Verlag GmbH, Kaufingerstraße 24, 80331 München, Germany

www.ingramcontent.com/pod-product-compliance
Lightning Source LLC
LaVergne TN
LVHW080655011026
849160LV00014B/1438
* 9 7 8 1 0 3 2 9 2 9 4 3 9 *